Elmostafa Gagou

Mycorrhizae: The future of agriculture in the Figuig oasis

Elmostafa Gagou

Mycorrhizae: The future of agriculture in the Figuig oasis

The rise of a new green revolution

ScienciaScripts

Imprint
Any brand names and product names mentioned in this book are subject to trademark, brand or patent protection and are trademarks or registered trademarks of their respective holders. The use of brand names, product names, common names, trade names, product descriptions etc. even without a particular marking in this work is in no way to be construed to mean that such names may be regarded as unrestricted in respect of trademark and brand protection legislation and could thus be used by anyone.

Cover image: www.ingimage.com

This book is a translation from the original published under ISBN 978-620-6-72756-9.

Publisher:
Sciencia Scripts
is a trademark of
Dodo Books Indian Ocean Ltd. and OmniScriptum S.R.L publishing group

120 High Road, East Finchley, London, N2 9ED, United Kingdom
Str. Armeneasca 28/1, office 1, Chisinau MD-2012, Republic of Moldova, Europe
Managing Directors: Ieva Konstantinova, Victoria Ursu
info@omniscriptum.com

Printed at: see last page
ISBN: 978-620-8-34880-9

TABLE OF CONTENTS

GENERAL INTRODUCTION

Due to the cumulative effects of climate change, date palm (Phoenix dactylifera L.) oases in North Africa are facing major challenges that are threatening their sustainability [1] and causing soil salinisation [2]. Despite these obstacles, date palm remains of considerable economic, ecological and social importance in the region. It is a crucial source of income for oasis producers [3] and creates a favourable microclimate for agriculture in arid regions [4]. The date palm is a fruit product of global importance, not least because of its ability to mitigate global warming and absorb carbon dioxide more effectively than other trees [5]. In addition, dates, the fruit of the date palm, are considered an ideal food, providing a wide range of essential nutrients and many potential health benefits [6,7].In Morocco, an area of 59,600 ha is dedicated to date palm cultivation, which is found in thirteen provinces in the south and south-east. Figuig, Errachidia, Tinghir, Ouarzazate, Tata, Zagora and Guelmim account for 98% of Morocco's palm grove. The Figuig oasis, located on the border between Morocco and Algeria, is part of the Oriental region of Morocco. Its palm grove covers an area of around 1,200 hectares, with around 260,000 date palms of ten different varieties [8]. Unfortunately, the palm trees in the Figuig oasis are vulnerable to "Bayoud", a disease caused by a fungus of telluric origin, Fusarium oxysporum f.sp. albedinis [9]. Every year, this disease decimates 4.3% of the date palm population in the oasis [10]. The fight against Bayoud is largely based on quarantine measures. Soil disinfection is costly and difficult, and chemical control is only possible if infection is detected early in a healthy area. Genetic control methods are also being studied, including the selection of resistant cultivars from natural date palm populations [11-13]. Despite the many rehabilitation programmes launched to remedy the critical situation of palm groves [14,15] , these initiatives remain incomplete, as any ecosystem restoration programme must incorporate components from different levels of the ecosystem, in particular soil micro-organisms that are closely linked to plants. The use of symbiotic microorganisms, such as arbuscular mycorrhizal fungi (AMF), has often improved the growth and tolerance of crops under various stresses [16-18].

As multifunctional symbiotic agents, these micro-organisms play a crucial role in improving various aspects of plant performance. According to studies, MCAs facilitate the uptake of nutrients and water by plants, thereby improving their growth [19-21]. They can also improve soil fertility [22] and enhance host plant tolerance to drought and salinity [17,23].

As a result, choosing the appropriate CMAs that enhance plant resilience is crucial for successful land rehabilitation. AMCs native to a specific soil have been shown to perform better than exogenous AMCs, as demonstrated by Estrada et al. (2013) [24]. It is therefore essential to analyse the mycorrhizogenic capacity of the rhizospheric soils of the Figuig palm grove in correlation with edaphic characteristics, with a view to potential exploitation of their AMC resources. It is also important to study the diversity of MCAs in the oasis, to develop a local collection of mycorrhizal strains and to set up an inoculum production unit with a view to using them biotechnologically to improve the health of palm trees and the soils on which they grow. Symbiosis between MACs and terrestrial plants is a common association, affecting 70-90% of plant species [25]. In the past, the taxonomy of these fungi was based mainly on spore morphology. Families and genera were distinguished on the basis of different modes of spore formation, while species were differentiated by spore-specific characteristics such as colour, size, subcellular structures and the phenotypic and histochemical properties of spore wall components [26-28] . Currently, molecular phylogenetics based on ribosomal DNA sequences is the only method for identifying the different MAC species. Although the phylum Glomeromycota is considered to be a small group of around 250 species based on morphological criteria [27,29] , it is clear that these fungi have no host specificity, although some species may be specialists, existing only in particular ecosystems and only with plants adapted to these environments.In the light of these data, we have focused this study on the following aspects: Assessing the mycorrhizal profile of rhizospheric soils of date palms grown in the Figuig oasis in correlation with edaphic parameters.To improve our understanding of the biodiversity of the Figuig oasis by identifying (morphologically) the strains of CMA isolated from the rhizosphere of date palms in the Figuig oasis and then establishing a referenced collection.

CHAPTER 1

THE FIGUIG PALM GROVE

The Figuig palm grove is located at the south-eastern tip of Morocco, more precisely in the Oriental region, directly adjacent to the border with Algeria (Figure 1). The area is characterised by its strategic position at the junction of the eastern High Atlas and the Saharan Atlas, making it one of the continental oases of the pre-Sahara. Topographically, the palm grove extends over two clearly differentiated areas: a plateau to the north and a plain to the south, separated by an escarpment known as the jorf, which is approximately 30 metres high. The climate in Figuig can be described as arid Mediterranean. The average annual temperature in the region is 20°C [30]. Large seasonal variations and daily amplitudes are included in this average. The average annual rainfall in Figuig is 122 mm, and the rainy period lasts from October to January. There are around 20 to 30 rainy days a year. The main hydrological resources of the oasis are the groundwater and the wadis. The land is no longer used for agricultural purposes to the same extent as it used to be, and there are several reasons for this change and reduction in use, such as the decline in the number of springs and their capacity, the increase in groundwater salinity and vascular fusariosis (Bayoud) of date palms [10,31].

1. Fusarium head blight of date palms: the main threat to phoeniculture in Morocco :

Another major threat to the biodiversity of date palms is the appearance over the last hundred years or so of a telluric disease of the tracheomycosis family, vascular fusariosis, or Fusarium head blight. commonly known as "Bayoud", caused by the fungus Fusarium oxysporum f. sp. albedinis (Foa) [32] . This disease has already caused substantial losses (Figure 2). It is estimated that more than two-thirds of palm trees have disappeared since the disease appeared. Moreover, the most susceptible varieties are cultivars of good date quality [33].

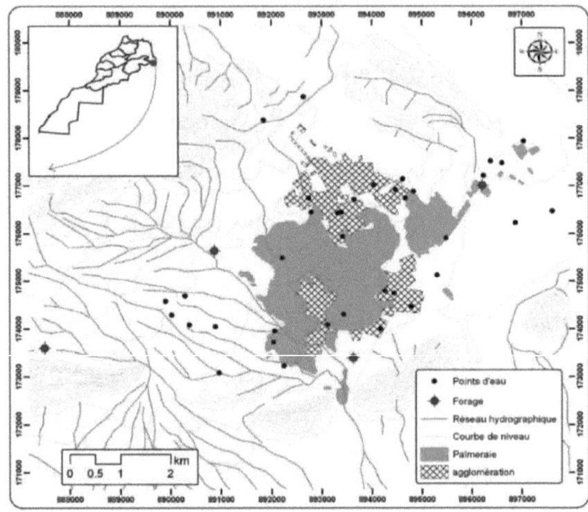

Figure 1: Geographical location of the Figuig oasis (From Jilali.A, 2015)[56].

Foa develops inside the xylem of the conducting vessels of palm trees [33,34]. It enters its host through the roots and remains present in the soil for many years (up to 8 years) thanks to the development of particularly resistant chlamydospores [33]. These remain trapped in the dead tissues of infected date palms until they decompose and are released. They can also remain inside healthy carriers such as henna, alfalfa or clover [35]. A A very limited number of spores are needed to cause the disease, and these have been found at depths of up to 1 m. This makes the application of fungicides particularly ineffective, as they are unlikely to reach this depth in sufficient quantities. This makes the application of fungicides particularly ineffective, as they are unlikely to reach this depth in sufficient quantity [33]. Moreover, this practice is highly incompatible with the oasis ecosystem and its fragile balance [36].

Figure 2: Date palm attacked by Fusarium head blight. The palms show characteristic symptoms of the disease, such as wilting and progressive drying.

The symptoms are both internal and external, although it is naturally the latter type that is visible first.

▪ **External symptoms**

• The first of these symptoms is discreet; it appears on a palm of the middle crown, which takes on a leaden hue.

• The leaflets on this leaf progressively fold to one side only, then the symptoms spread to the opposite side. The leaflets turn white from their basal part to their distal end (Figure 3).

• The palm dries out completely and ends up hanging down the stipe.

• The same symptoms are affecting neighbouring palms.

• Infected palms may die within a few weeks to a few months of infection, depending on the variety and cultivation practices. This occurs when the parasitic fungus reaches the palm's terminal bud.

• At root level, infection is very localised. Only a small number of roots turn reddish-brown and the tissues decompose. This is actually the first symptom, but it can generally only be observed when the tree is uprooted once it is dead [33].

Figure 3 : Symptom of Bayoud visible on half of a date palm.

- **Internal symptoms**

Internal symptoms can be observed by making transverse and longitudinal cuts through the various infected parts of the palm. At the base of the stipe, corresponding to the small number of infected roots, there is a small group of cribro-vascular vessels which, together with the surrounding parenchyma and sclerenchyma, take on a mahogany or mustard colour. In transverse sections, they form patches measuring a few square centimetres, while in longitudinal sections they correspond to elongated patches. Further up the stipe, the infected conductor bundles become divide and it is possible to follow the path of colonisation of the different palms affected by the parasitic fungus [37].

To date, there is no curative method and infected palms are systematically burnt on site once the disease has been identified [33]. The most widespread control method consists of encouraging the use of resistant varieties to the detriment of more susceptible varieties, which are gradually being lost [38]. Several studies have already identified resistant cultivars such as Bousthammi-noire, Tadment, Boukhanni and Bousthammi-blanche, varieties that are strictly resistant to the disease [38,39]. Recently, it was shown that the use of compost made from the organic waste of date palms in the growing medium reduced the mortality of seedlings exposed to Foa [36]. This observation can be explained, on the one hand, by the fact that compost contains a highly diversified microbial flora that competes with Foa and has several microorganisms that are antagonistic to Foa and, on the other hand, by the presence of lignocellulosic substances derived from the decomposition of date palm organic waste [36]. However, the genetic determinism of resistance has not yet been explained [38].

2. The drought

Despite date palms' adaptation to arid climates, droughts are a real threat to palm groves. These extreme conditions can cause considerable damage and even lead to the complete disappearance of palm groves. In the 1980s, a period marked by a particularly intense drought, Morocco suffered the loss of no fewer than 350,000 plants [40]. This period of low rainfall had disastrous consequences for water resources, seriously impacting aquifers and, by extension, the health of palm groves. In addition to the losses Although these droughts have a considerable economic impact, they also lead to a loss of varietal biodiversity. Certain varieties that are less resistant to lack of water are the result of millennia of selection by plant breeders and represent a reservoir of genetic variability that is essential for future improvements [40]. Preserving this varietal diversity is therefore crucial. All the more so as the intensity and frequency of droughts are likely to worsen in the years to come, due to global warming [41,42]. These prospects underline the urgent need to develop adaptation and conservation strategies to protect these vital ecosystems and the economies that depend on them.

3. The salinity of irrigation water

Many countries, particularly those located in arid and semi-arid regions, are faced with the salinisation of soils (Figure 4) and water resources [43]. Natural water used for irrigation inevitably contains solubilised mineral salts, which are produced by the interaction of water with the rocky materials or solid deposits through which it passes. The ions most frequently found in these natural solutions include chlorides, sulphates and bicarbonates of calcium, magnesium and sodium. The quality and suitability of irrigation water is determined by the particular concentration and composition of these mineral salts [44].

The Saharan regions are mainly irrigated by groundwater from aquifers with a high salt content. Dry residue levels are frequently higher than 4 to 5 g/l, and can sometimes be as much as three times higher [45]. However, it should be pointed out that in certain situations, the salt content of Saharan water can be as low as 0.5 g/l, which could be considered adequate for human consumption. Salt concentrations of up to 2 g/l classify the water as being of excellent quality for irrigation. However, water with a salinity of between 2 and 5 g/l is considered salty, and water with a concentration of more than 5 g/l is classified as very salty [46]. The salinity tolerance threshold for irrigation water is generally set at 3 dS/m, which corresponds approximately to 2 g/l [47]. However, the high evapotranspiration associated with drought conditions means that even this salt concentration can cause significant problems for plants, particularly those grown in fine-textured soils or under intermittent irrigation [48]. In terms of the ionic composition of Saharan waters, sodium often predominates, generally accounting for half of the total cations [46]. As for anions, chlorides and sulphates are the most widespread.

Figure 4: Saline crusts in the Touteline palm grove (Tata Province) - Study by CNEARC, ALCESDAM, DPA (2004)

4. Deterioration of the oasis soil

In general, soils in arid regions are characterised by a lack of water, low fertility, a sandy texture, and accumulation layers rich in salts, limestone and gypsum. The deterioration of soils in oasis zones represents a significant threat to the sustainability of these ecosystems. The impacts of this drop in productivity, amplified by climate change, rising temperatures and extreme weather events, seriously compromise the future of agriculture and environmental stability in these areas [49]. In addition, the degradation of oasis soils is leading to a reduction in biodiversity, disrupting the activity of fauna, microfauna and soil microbiota. These disturbances have a major impact on the environment, favouring droughts, fires, desertification and salinisation [49,50]. Degradation also negatively affects the metabolism of micro-organisms and plants, destroying certain layers of the primary food chain in the soil [50]. Degradation of oasis soils affects biodiversity in several ways:

• **Loss of soil biodiversity**: Soil degradation, caused by factors such as intensive agriculture, deforestation and overgrazing, leads to a loss of soil biodiversity. This is detrimental to the fertility of ecosystems and agricultural production [49,51].

- **Alteration of surface ecosystems**: Soil degradation affects surface ecosystems by altering decomposition levels and nutrient retention, and by leading to a reduction in plant cover [51].

- **Impact on fauna and microfauna**: erosion, soil compaction, reduction in organic matter, salinity and pollution. disrupt the activity of soil fauna, microfauna and bacteria, leading to a loss of biological diversity [49].

- **Water quality problems**: The physical and chemical degradation of soils, such as erosion and pollution, also affects water quality, particularly in terms of solid load and pesticide pollution [49].

To reduce soil degradation and preserve biodiversity, the adoption of environmentally-friendly farming practices, such as organic farming, is crucial. These sustainable methods are essential not only for preserving soil biodiversity, but also for responding effectively to current environmental challenges. This approach could also highlight the potential role of arbuscular mycorrhizal fungi in the resilience of oasis soils, helping to integrate oasis issues into national sectoral policies [52].

5. The Figuig oasis

The Figuig oasis is facing significant soil degradation, intensified by a combination of factors. These factors include the irrational exploitation of water resources, the adoption of unsuitable farming practices and difficult environmental conditions. In addition, the concentration of the local population around the oases and an economy heavily dependent on traditional agriculture, which is vulnerable to fluctuations in water availability, have amplified this deterioration. It is also essential to mitigate soil degradation and revitalise oasis systems. These actions are crucial not only for preserving the region's agricultural biodiversity, but also for improving food security [49,52,53]. The main causes of soil degradation in the Figuig oasis include :

• **Water scarcity and quality:** In desert areas such as the Figuig oasis, persistent water scarcity and deteriorating water quality pose serious challenges. These problems make irrigation difficult and contribute to soil salinisation [52,54].

• **Unsuitable cultivation practices:** Excessive irrigation, particularly of date palms, has aggravated salinisation and soil degradation. These effects are intensified by the impact of irrigation on shallow water tables [52].

• **Difficult environmental conditions and low soil fertility:** The oasis ecosystems of Figuig face difficult environmental conditions and naturally low soil fertility, contributing to their ongoing degradation [49,50,52].

Faced with these challenges, it is crucial to focus on the sustainable management of water resources, the adoption of appropriate irrigation practices and the implementation of strategies to improve soil quality and fertility, taking into account the specific environmental conditions of the region.

CHAPTER 2

ARBUSCULAR MYCORRHIZAL MUSHROOMS

1. The importance of fungi in ecosystems

The majority of biological entities in the soil matrix are micro-organisms, which represent a significant proportion of the earth's genetic diversity. Because of their great diversity and widespread presence, fungi play an essential role in terrestrial ecosystems. It is estimated that one gram of soil can harbour between 1010 and 1011 fungi [55], between 6,000 and 50,000 distinct bacterial species [56], and contain up to 200 metres offungal hyphae [57]. The key role attributed to fungi is due to the influences they can have on the chemical processes that govern ecosystems, such as the provision of essential nutrients for plants [25], as well as on major geochemical cycles, notably the nitrogen cycle [58] and the carbon cycle [59].

6. General information on mushrooms

2.1 Phylogenetic position of fungi

Because of their common morphological characteristics, fungi were previously grouped together in the kingdom of Plants. The fungal kingdom, also known as Mycota, dates back to 1969 [60]. It was based on specific characteristics, such as the absence of chlorophyll or starch. The current name of this association, "Fungi", derives from the Latin term "Fungus", which means "fungus" [61]. Using structural classification approaches and molecular methods, it has been suggested that fungi probably share a common ancestor with the animal kingdom [62]. However, these two kingdoms were finally recognised as brothers [63,64]. Fungi, like animals, are included in the Opisthocontes group of the Empire of Eukaryotes [64].

2.2 Mushroom characteristics

Fungi are eukaryotic organisms that can exist as unicellular or multicellular entities. They are recognised by their cell walls, which are composed of polysaccharides, hemicellulose and chitin [65]. Fungi are incapable of photosynthesis due to the absence of chlorophyll and plastids. The vegetative organs of fungi are called hyphae and are composed of thallus. These hyphae may or may not be septate and facilitate the process of absorbing nutrients by osmosis. Fungi have a sexual or asexual life cycle. There are currently around 99,000 species classified in the fungal kingdom, but estimates suggest that there could be as many as five million species of fungi [66]. These species are classified into eight classes and one sub-range (see Figure 5). The sub-region Dikarya [67] encompasses the phyla Basidiomycota and Ascomycota, while other phyla such as Glomeromycota, "Zygomycetes", Chytridiomycota, Blastocladiomycota, Neocallimastigomycota, Microsporidia, Olpidium and Rozella are also included. However, this classification is still under development and discussion.

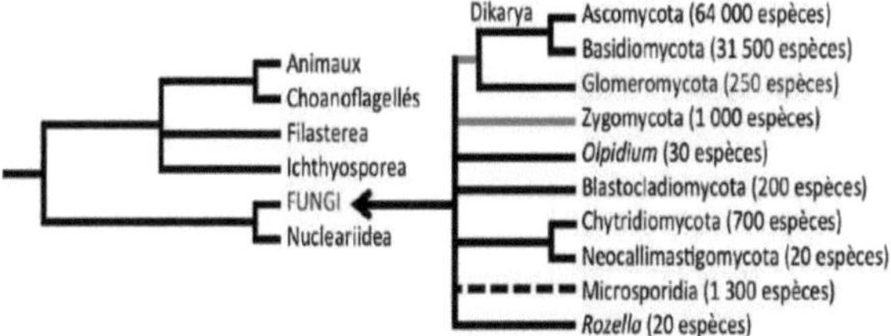

Figure 5: Revised phylogeny of Opisthocontes according to Keeling et al. (2009) and Blackwell et al. (2012) [87]: grey branches for uncertain monophyly and dotted lines for an unconfirmed phylogenetic position.

2.3 How mushrooms live

Fungi are considered to be chemoheterotrophs, which means that they draw their energy from carbon molecules in the environment, as they cannot create them. They have different nutrient methods, and therefore different lifestyles. This variety stems from the fact that they are common to both terrestrial and aquatic ecosystems. Some fungi are considered to be parasites and cause diseases that are sometimes fatal to the host. These parasites attack plants, fungi, algae, animals and even humans, such as aspergillosis caused by an ascomycete of the genus Aspergillus. Some fungal species, such as the genus Malassezia, are considered to be commensal, meaning that they benefit from their host without having a negative impact on it, but they are also considered to be parasitic.Many species of fungi are saprotrophic, which means that they feed on the degradation of dead organic material such as wood, leaves and corpses. The fungus Tricholomopsis ornata degrades dead leaves. Finally, the most recent way of life observed in fungi is symbiosis, which is a lasting relationship with another organism. This type of interaction is created by fungi with different partners. For example, lichens are symbiotic groups of fungi and photoautotrophs, such as cyanobacteria and green algae [68]. Plant roots and fungi can also form this mutualistic symbiosis. In this case, the symbiont is defined as a mycorrhiza.

2.4 The different mycorrhizal symbioses

There are generally four main types of mycorrhizae (Table 1): arbuscular mycorrhizae, ectomycorrhizae, orchid mycorrhizae and ericoid mycorrhizae [69].

Table 1: Characteristics of the main types of mycorrhizal symbiosis

Type	Structures trained	Plants concerned	Mushrooms involved	Breakdown geographical
Arbuscular endomycorrhiza (AM)	Arbuscules and intracellular vesicles	Bryophytes, Pteridophytes, Gymnosperms and Angiosperms, around 200,000 species	Glomeromycetes (Glomales)	Everywhere, especially in tropical regions
Endomycorrhiza of orchids (ORM)	Intracellular bundles	Orchidaceae, around 25,000 species	Basidiomycetes (Rhizoctonias) and Ascomycetes (rarely)	Everywhere, in every orchid biome
Endomycorrhiza of ericaceae (ERM)	Intracellular bundles	Ericaceae, about 1 500 species	Ascomycetes (Helotiales)	Everywhere, in every ericaceous biome
Ectomycorrhiza (ECM)	Thickened root, mantle and Hartig network	Gymnosperms and Angiosperms, around 10,000 species	Basidiomycetes and Ascomycetes	In temperate regions where it is dominant

7. Arbuscular mycorrhizal fungi

3.1 Introduction

Arbuscular mycorrhizal fungi (AMF) are soil fungi distributed worldwide, forming symbioses with around 72% of terrestrial vascular plants, including the vast majority of agricultural and horticultural crops [70]. These fungi have been found in plant fossils dating back 460 million years, suggesting their role in the colonisation of terrestrial ecosystems by aquatic plants [71]. They are now classified in a monophyletic phylum, the Glomeromycota [72,73].

In the CMA symbiosis (Figure 6), the fungus develops an intraradical mycelium (IRM) inside the roots and an extraradicular mycelium (EMM) in the soil. These two components are interconnected to form a vast mycelial network composed of innumerable coenocytic hyphae, i.e. without separation into compartments [74], and containing hundreds of nuclei sharing the same cytoplasm.

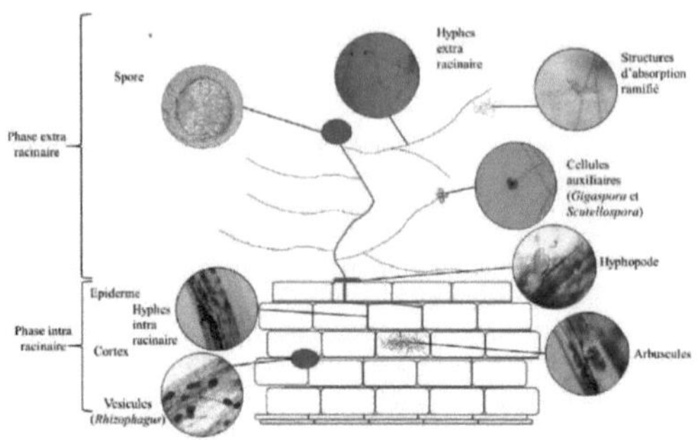

Figure 6: Structures of arbuscular mycorrhizal fungi

In this association with plants, the fungus receives photosynthetic compounds, such as lipids and carbon, in exchange for water and minerals, mainly phosphorus [25,75]. This symbiosis often results in an improvement in the biomass and yield of plants, as well as an increase in their resistance and tolerance to biotic and abiotic stresses.

3.2 Classification of arbuscular mycorrhizal fungi

Morphological and phylogenetic studies have made it possible to classify all arbuscular mycorrhizal species, belonging to the phylum Glomeromycetes, as illustrated in Figure 7 [76]. To date, there are four main orders of mycorrhizal fungi: Glomerales, Paraglomerales, Archaeosporales and Diversisporales [73,77]. The Glomeromycetes are characterised by considerable morphological diversity, particularly in the spores, which vary considerably in size, colour and shape depending on the species. There are currently around 18 genera and 250 species of Glomeromycetes. Their current classification is based mainly on the isolation of species from spores. However, the use of modern molecular techniques, which allow genetic sequences to be extracted from roots or soil, makes it easier to identify species that are non-sporogenic or otherwise difficult to detect. The real number of species could therefore be much higher than the current estimate [78].

3.3 CMA life cycle, structures and functions

3.3.1 CMA life cycle

CMAs are obligate symbionts that depend on a photosynthetic host for their survival and to complete their life cycle. The life cycle comprises three key phases: establishment of symbiosis, vegetative growth and dispersal. The initial phase of symbiosis establishment begins with host recognition and the actual establishment of symbiosis (Figure 8).

This recognition involves the activation of soil propagules, leading to the spontaneous germination of a spore and the formation of germinative hyphae. In the absence of host recognition, germination may slow down or stop, while retaining sufficient carbon resources for further germination. Establishing symbiosis requires effective communication between the two partners in the soil.

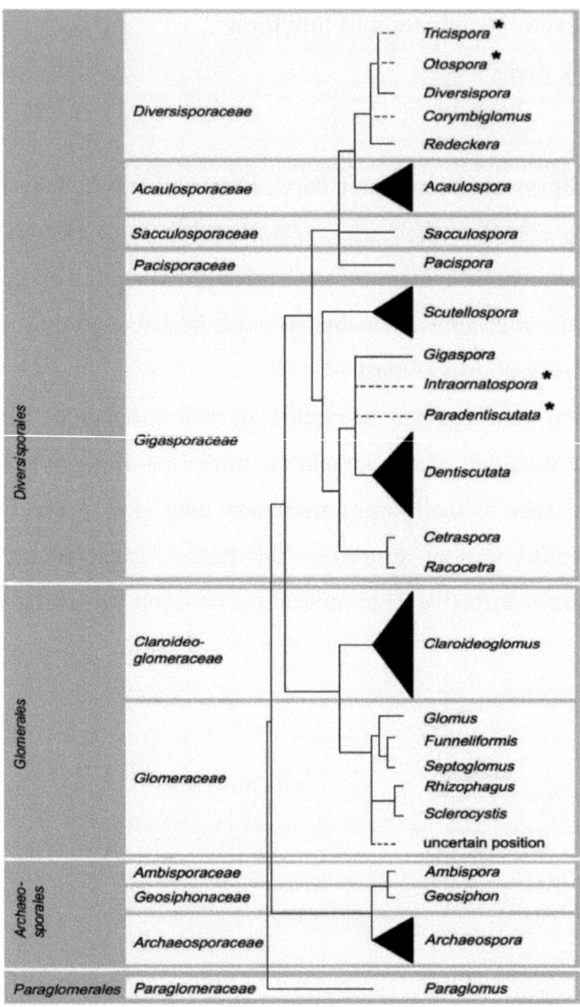

Figure 7: Classification of arbuscular mycorrhizal fungi (after Redecker et al.al.,2013)[106]

The intensification of germinative hyphal formation results from the exchange of molecular signals. In response to root exudates, in particular strigolactones secreted by a host plant [79,80], the hyphae modify their morphology, adopting an intense branched growth [81]. The germinative hyphae develop and, by branching, form a pre-symbiotic mycelial network directed towards the root. At the same time, the CMA releases "Myc" factors into the soil, stimulating activation of the symbiosis signalling pathway in the plant [82], which encourages the formation of more lateral roots [83,84]. On contact with the root, the CMA forms an appressorium, a structure specialised in the recognition and interaction between the two partners, marking the start of the fungus's penetration of the root. During the establishment phase of the symbiosis, the fungus develops intraradicularly, forming intercellular hyphae. Once it reaches the cortical cells of the root, it creates highly branched structures within the cells: arbuscules. These arbuscules, which give their name to this type of symbiosis, are essential for the exchange of nutrients between the two organisms. Their formation induces significant morphological changes in the host cell, without compromising its integrity. At this stage, the symbiosis is fully established.The MAC life cycle then continues with the vegetative phase. During this stage, the mycelium develops intensively outside the root (extraradicularly). Some MACs also form intraradicular storage structures, vesicles, which may be located inside or between the cells. This phase is marked by an increase in fungal biomass, thanks to the extension of the mycelial network, which explores the soil and can potentially colonise other plants.The cycle ends with the reproduction phase, characterised in CMAs by sporulation. This stage consists of forming propagules, mainly spores, which allow the species to spread and perpetuate itself.

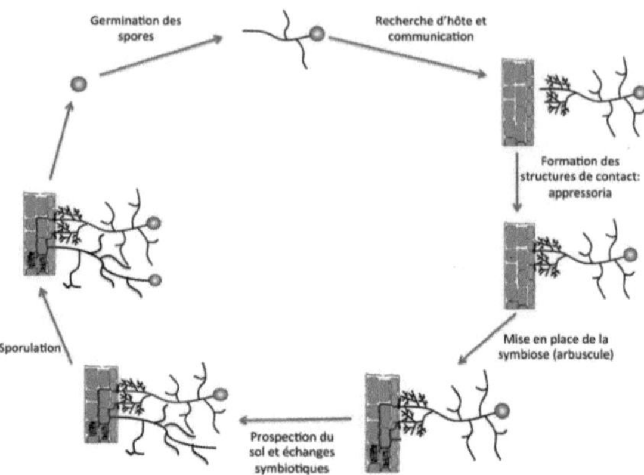

Figure 8: Diagram of the life cycle of a CMA, (Adapted from Besserer, 2008)[112].

CHAPTER 3

ASSESSMENT OF THE MYCORRHIZAL POTENTIAL OF SOILS IN THE RHIZOSPHERE OF DATE PALMS (PHOENIX DACTYLIFERA L.) IN THE FIGUIG OASIS (SOUTH-EAST MOROCCO)

1. Introduction

The date palm (Phoenix dactylifera L.) is recognised as an essential fruit crop worldwide, offering distinct environmental and nutritional benefits. This tree species plays a promising role in the fight against global warming thanks to its effective ability to sequester carbon dioxide, surpassing the capacities of many other trees [85]. In addition, the fruit (the date) is rich in essential nutrients, making it an ideal food choice, with potentially significant health benefits [7,86]. In the Middle East and North Africa, date palms play an essential role, not only ecologically but also economically, as they are an important source of income for oasis farmers [3]. Moreover, date palm cultivation favours a favourable microclimate, making agriculture possible even in the harsh conditions of the desert [4]. Despite their importance, date palm oases face a number of challenges that threaten their sustainability. Climate change has imposed cumulative effects and soil salinisation, caused by factors such as limited water resources, increased salinity of irrigation water and reduced drainage of agricultural land, is emerging as a pressing threat to these oases [1,87-90]. The Figuig oasis in south-east Morocco is a good illustration of these challenges. This region is facing major threats that affect both its agricultural efficiency and its biodiversity. Dwindling water resources, a This problem is exacerbated by climate change, and the increase in soil salinity poses considerable challenges. The situation is further complicated by the presence of fusariosis, a disease caused by Fusarium oxysporum f. sp. albedinis (Foa)

[39,91-93].Faced with these challenges, and with the aim of mitigating the harmful effects of biotic and abiotic stresses, it is essential to direct research towards alternative approaches such as the use of soil micro-organisms that can promote the growth of date palms in these fragile environments. Among these micro-organisms, arbuscular mycorrhizal fungi (AMF), which establish a symbiosis with the majority of plants grown in agriculture and horticulture, are of particular interest. These fungi offer numerous advantages for plant growth and health in stressful environments [94,95]. In arid conditions, for example, plants colonised by MACs have shown greater tolerance to drought [96] and better access to phosphorus than non-mycorrhised plants [97]. CMAs can also improve the stability of soil aggregates [98], which is crucial in erosion-prone sandy soils. In extreme desert ecosystems, MACs play a key role in the development of vegetation. For example, inoculation with MACs has been shown to improve water and nutrient uptake in desert succulents [99].Research by Meddich.A et al (2015) [100] revealed the important role of native CMAs isolated from the Aoufous palm grove in the tolerance of date palms to water deficit and fusarium head blight. These CMAs also serve as bioindicators, as the characteristics of agricultural soils can be determined on the basis of their mycorrhizal fungal communities. Despite their interest and importance, CMAs are rarely used on farms, partly because of the incompatibility of their use. of the introduced isolate with the characteristics of the local soil [101], leading to the disappearance of the introduced inoculum. It would therefore be wise to select indigenous isolates adapted to the constraints of the Moroccan oasis environment. Indeed, the adaptation of MAC inoculums to specific environmental conditions has been widely documented [102-104]· It has been shown that MACs perform best when experimental conditions most closely resemble those of their native environment [104]. It can therefore be assumed that AMCs isolated from desert ecosystems are better adapted to cope with the prevailing stress conditions and may exhibit unique physiological capabilities.

In this context, the aim of our work was to study the soils of the Figuig palm grove and assess their mycorrhizal status, with a view to their potential use. It is essential to characterise CMA isolates adapted to date palm ecosystems, as a prerequisite for future fundamental and applied research projects.

2. Study sites and sampling

The Figuig oasis is made up of several ksour: Zenaga (ZG), Oudaghir (OD), Lamaiz (LZ), Ouled Slimane (OS), Laabidate (LB), and Elhammam (EH), which were originally separate settlements but have since been merged due to urban expansion. The topography of the region is characterised by two distinct sections, separated by the escarpment known as the jorf. The largest ksar, Zenaga, lies in the lower section, while the other five ksour and their associated palm groves are located in the upper section, at an altitude of 899 metres above sea level. On the outskirts of these urban areas are extensive palm groves, also known as Extension (EX) and Aarja (AR), as shown in Figure 1. For a full survey of the palm grove, eight different sites have been carefully selected, each representing a specific palm grove

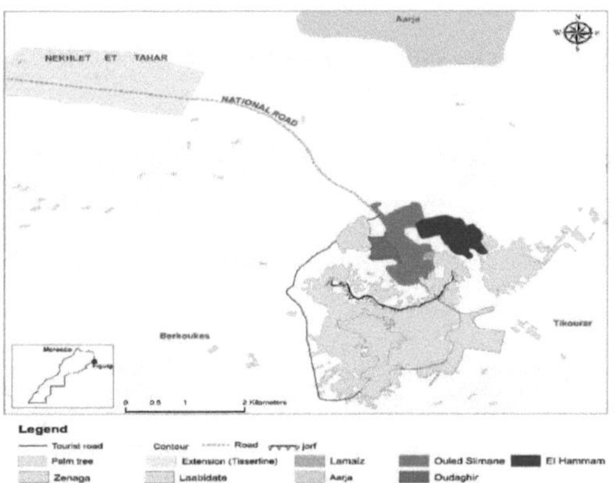

Figure 1: Satellite view of the Figuig oasis with indication of sampling sites samples

Soil samples were taken at each site. Three replicate plots were randomly selected per site, each covering an area of 700 m^2. In each plot, seven date palms were carefully selected, and four soil sub-samples were taken around each tree, at a depth of 0-30 cm and at a distance of 50 cm from the trunk. These sub-samples were combined to create one composite sample per plot, resulting in a total of 24 composite samples for further analysis.

3. Physico-chemical analysis of surveyed soils

3.1 Soil pH

Soil pH analysis was carried out according to the method of Eaton et al (2005) [105], using a Biobase China pH-920 pH meter calibrated with standard buffer solutions to ensure accuracy. Initially, an aqueous soil extract was prepared by mixing 10 of 2 mm sieved soil with 50 ml of distilled water. To ensure complete homogenisation, the suspension was stirred for 30 minutes, allowing the pH of the soil in the suspension to equilibrate. After a rest period of 15 minutes, the glass electrode of the pH meter was immersed in the extract to measure the pH.

3.2 Assimilable phosphorus content of soils

The Olsen method, used to determine the presence of phosphorus in soil, is based on the formation of a complex between orthophosphoric acid and molybdic acid, as described by Olsen et al. (1954) [106]. After phosphorus has been extracted from soil samples, this reaction takes place in the presence of a 0.5 M sodium bicarbonate solution ($NaHCO_3$) at pH 8.5. The reaction between the phosphorus in the sample and the molybdic acid forms a phosphorus-molybdate complex. This complex is then reduced, giving rise to a characteristic sky-blue colour, the intensity of which is directly proportional to the amount of phosphorus present in the sample. The optical density (OD) of this coloured solution is measured using a spectrophotometer set at 820 nm to assess the phosphorus concentration. By converting the intensity of the colour into a measurable concentration, this spectrophotometric method provides an accurate estimate of the phosphorus content available in the soil.

3.3 Soil organic matter content

The method of Walkley and Black (1934) [107], used to measure the quantity of organic matter (OM) in the soil, involves the oxidation of organic matter in an acid medium using a potassium dichromate solution. For this process, a 1 gram soil sample is treated with a reactive mixture of potassium dichromate and concentrated sulphuric acid, inducing rapid oxidation of the organic matter. After this reaction, a solution of ammoniacal ferrous sulphate (0.5 N) is used to titrate the remaining unreduced potassium dichromate. The formula for calculating the quantity of organic matter is as follows:

$$\%MO = \frac{(Vb - Vs) \times N \times 0.3}{P} \times 100$$

- % OM represents the percentage of organic matter in the soil sample.
- Vb is the volume of titration solution used for the blank.
- Vs is the volume of titration solution used for the soil sample.
- N is the normality of the potassium dichromate solution.
- 0.3 is a conversion factor corresponding to the quantity of oxygen consumed by 1 millilitre of standardised dichromate solution.
- P is the weight of the dry soil sample in grams.

3.4 Potassium content of soils

The 1 M ammonium acetate extraction method described by Mathieu and Pieltain (2003) [108] was used to assess the potassium content of the soil. To extract potassium (K) efficiently, sieved soil samples are mixed with this solution and agitated. For agronomic purposes, the potassium concentration is measured after filtration, often in the form of potassium oxide (K_2O).

3.5 Determining soil salinity

Following the method described by He et al. (2012) [109], the electrical conductivity (EC) of a soil-water extract was measured to determine soil salinity. To do this, a soil sample is first mixed in a 1:5 ratio with distilled water. The extract is filtered after a sufficiently long resting period to allow the soluble salts to dissolve. A conductivity meter is then used to measure the electrical conductivity of the extract.

3.6 Calcium carbonate content of soils

The calcium carbonate ($CaCO_3$) content of the soil was determined using a Bernard calcimeter, in accordance with the French standard NF P 94-048 [110]. In this method, a soil sample is treated with an acid, usually hydrochloric acid, inside the calcimeter. The reaction of the acid with the soil carbonates generates carbon dioxide (CO_2), the volume of which is measured by the calcimeter. This volume of CO_2, using known stoichiometric relationships, is then used to calculate the calcium carbonate content of the soil.

3.7 Determining soil texture

According to Ritchey et al. (2015) [111], soil texture was determined using the tactile method, also known as the 'Feel Method', to assess soil texture (Table 2). This technique involves physically manipulating a wet soil sample to determine its texture class - sand, silt or clay. By rolling it between the fingers and forming it into ribbons or balls, the analyst can assess the cohesion, smoothness and plasticity of the sample. Sandy soils, which are granular in nature, do not form ribbons, whereas clay soils, which are characterised by high plasticity, can form long ribbons. Silty soils fall between these two extremes. Although subjective, this method is widely recognised for its usefulness in the rapid assessment of soil texture classes in the field. It provides a better understanding of the physical properties of the soil, such as water retention, permeability and bearing capacity.

3.8 Assessment of the number of MAC infectious propagules in the soils studied

The assessment of the number of MAC infectious propagules in the different soils was based on the Most Probable Number (MPN) method. This bioassay measures the presence or absence of MAC propagules (by observing root colonisation) in a series of soil dilutions, the results being interpreted as a probabilistic estimate of the number of propagules from a statistical table [112]. The test procedure began by air-drying the soil samples to remove moisture. The samples were then sieved to obtain a particle size homogeneity of 2 mm. This sieving process prepared the soil for subsequent handling. After sieving, the samples of soil were diluted with previously sterilised sand. Sterilisation was achieved by heating the sand to 180°C for three hours.

Table 2: Tactile test of soil texture (approximate method)

Soil texture	Dry soil	Wet soil
Sandy soils	The grains of sand are visible to the naked eye. The ground flows through your fingers like sugar. The ground is very granular and abrasive.	It's very difficult to mould the floor and it breaks when you touch it. The floor doesn't stick to your fingers; it's rough and abrasive to the touch.
Silty soils	The floor has a powdery or floury appearance. The floor is soft to the touch.	The ground is very soft and slippery like soap. You can form a ribbon with the earth by Roll it between your hands; the ribbon breaks if you try to bend it. The floor is not very sticky.
Clay soils	The soil is made up of very hard clods that are difficult to break up.	The floor is very sticky, smooth and shiny. The ground is very easy to shape; you can form long, flexible ribbons by rolling the soil between your hands.
Loamy soils	The soilis a little grainy. The floor can be handled carefully, without breaking the clods.	The floor is a bit sticky and grainy. If you roll the earth between your hands, you can form a ribbon, which cracks a little.

Sterilisation was essential to avoid contamination or the development of micro-organisms that could have falsified the test results. Each soil sample was subjected to six separate dilutions, with factors of 1/4, 1/16, 1/64, 1/256 and 1/1024. Each dilution was repeated five times to obtain statistically significant data and minimise experimental error. Plastic jars with a capacity of 200 ml were then used to contain the diluted soil samples. Each jar was filled with 100 g of diluted soil, representing an exact quantity of non-sterile soil from each dilution (Figure 10). Maize (Zea mays L.) was chosen as the symbiotic partner, due to its notable mycorrhizal dependency, its germination rate This is due to its considerable growth potential, its early receptivity to mycorrhizal colonisation and its prolific root production [113]. To achieve this, maize seeds were subjected to a surface sterilisation procedure, which involved immersing them in a 10% v/v sodium hypochlorite solution for 10 min. The seeds were then carefully rinsed with sterilised water to ensure effective sterilisation. After one week, each seedling was transplanted into the pot and carefully placed in the greenhouse, with precise control measures in place to maintain a constant temperature of 25°C and humidity of 80%. After a 30-day growth period, the plants were removed from their pots, and a root preparation procedure was followed. The roots were thoroughly washed and then treated with 10% KOH at 90°C for 15 minutes, followed by rinsing with 1% HCl for 10 minutes. They were then stained using blue ink at 90°C for 20 minutes, in accordance with the protocol established by Phillips and Hayman in 1970 [114]. This staining made the CMA structures in the roots visible. The roots were then cut into 1 cm segments and placed between a microscope slide and a coverslip for microscopic examination. A root system is said to be colonised by CMAs if it contains at least one infection point, indicating the penetration of hyphae into the root. The most probable number of propagules was determined using the following formula:

$$\log MPN\ (Most\ Probable\ Number) = (x \log a) - K$$

- x represents the average number of plants colonised by CMAs.
- a is the dilution factor
- K are listed in the tables published by Fisher et al (1949) [115].

Figure 2: Test for estimating the mycorrhizogenic infectious potential of surveyed soils

3.9 Assessment of spore numbers and identification of CMA species

In parallel with the MPN method, the direct extraction of spores from soils was carried out using the wet sieving method described by Gerdemann et al. (1963) [116]. This method requires a series of progressively smaller sieves (500 µm, 250 µm, 100 µm and 40 µm) (Figure 11). The material retained by the last three sieves was collected in 50 ml Falcon tubes and centrifuged for 2 minutes at 900 g in the presence of a 70% sucrose solution. The supernatant solution collected in the 40 µm sieve was carefully rinsed with tap water. The extracted spores were then distributed in Petri dishes. To determine the quantity of spores in each sample, five soil samples were examined under a stereomicroscope. The number of spores per gram of soil was used to define the number of spores found. In addition, the isolated spores were used to identify CMA species.

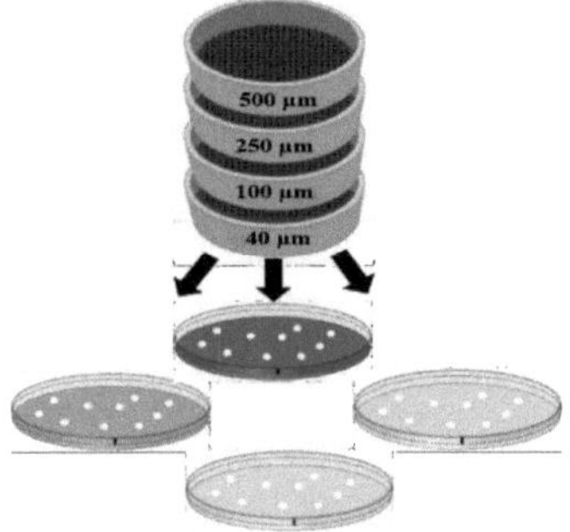

Figure 3: Diagram of the process for extracting CMA spores by wet sieving

Morphological characteristics of CMA spores, including colour, size, shape, number of walls, abundance, suspended hyphae and internal structure, were assessed using a stereomicroscope. Permanent samples created with a solution of polyvinyl alcohol, lactic acid and glycerol (PVLG), as described by Koske et al (1983) [117], and a mixture of PVLG and Melzer's reagent, as described by Brundrett et al (1994) [118], were used to test these characteristics (Figure 12).

Morphological characterisations provided by CMA Phylogeny [http://www.amf-phylogeny.com] [119] (accessed 17 May 2023) and the International Culture Collection of Vesicular Arbuscular Mycorrhizal Fungi [https://invam.ku.edu/species-descriptions] [26] (accessed 17 May 2023) were used to identify the spores.

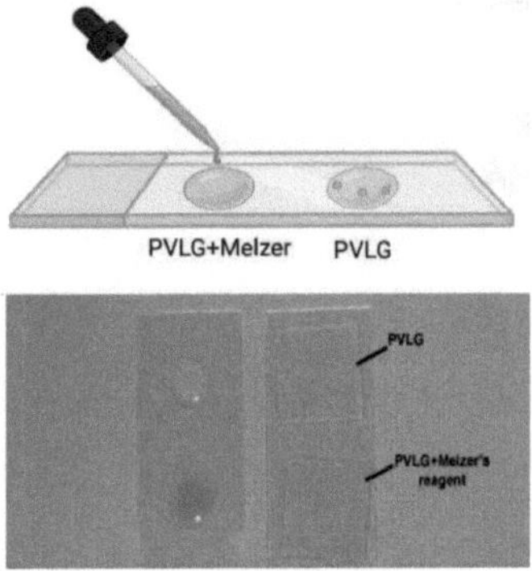

Figure 4: Preparation of diagnostic slides for MCA identification: PVLG with and without Melzer reagent

3.10 Mycorrhization frequency and colonisation intensity of maize roots

Using the method of Koske and Gemma (1989) [120], maize roots collected in the field were assessed for root colonisation by MACs. Briefly, the roots were cleaned in a 10% potassium hydroxide (KOH) solution at 90°C for 10 minutes before being stained at 70°C for 30 minutes with 2% Parker's blue ink (manufactured by Parker Inc., New York City, NY, USA) in 1% HCl prepared from a 37% concentrated solution. The rate of colonisation by MACs was calculated by placing 15 stained roots 1 cm long on glass slides and calculating the percentage of colonisation (Figure 13).

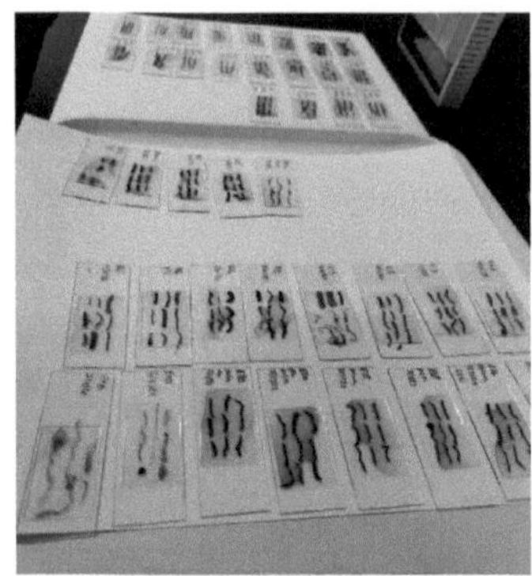

Figure 5: Preparation of maize root sections for microscopic observation

The rate of colonisation by MACs was calculated by placing 15 stained 1 cm long roots on glass slides and calculating the percentage of colonisation. The availability of MAC structures (hyphae, arbuscules or vesicles/spores) was determined by examining hyphae, arbuscules and vesicles/spores on 90 root fragments per site. According to Trouvelot et al (1986) [121], mycorrhizal colonisation of the root system was scored on an intensity basis from (0 to 5). A value of 0 indicates the absence of CMA colonisation (0%), while a score of 1 indicates the presence of minimal CMA structures (1%). Scores of 2, 3, 4 and 5 indicate progressively higher levels of MCA colonisation: 2 (1-10%), 3 (10-50%), 4 (50-90%) and 5 (greater than 90%), respectively. Mycocalc [https://www2.dijon.inrae.fr/mychintec/Mycocalc-prg/download.html][122] (consulted on 12 April 2023) was used to calculate the intensity (M%) and frequency (F%) of CMA colonisation in the root system:

• F% = (total number of mycorrhizal root fragments/observed root fragments) * 100

• M% = (95n5 + 70n4 + 30n3 + 5n2 + n1/total root fragments observed) * 100, where n5, n4, n3, n2, and n1 represent the total number of fragments classified as 5, 4, 3, 2 and 1 respectively.

4. Statistical analysis

The intensity and frequency of mycorrhizal colonisation, as well as the number of MAC spores, were analysed using single-factor assessments of variance (ANOVA 1), accompanied by Tukey's test at the 0.05 significance level. The Q-Q Plot was used to examine the normality of the residuals. The correlation between mycorrhizal symbiosis parameters and soil chemical analysis was analysed using Pearson correlation. SPSS statistical software (Version 21.0.0.0, 32-bit Edition) was used for the analysis (IBM SPSS Inc., Chicago, IL, USA).

5. Results and discussion

5.1 Physical and chemical properties of soil

The physico-chemical properties of the soil in the rhizosphere were carefully examined to determine their influence on the distribution and abundance of arbuscular mycorrhizal fungi (AMF). Soil texture results were obtained using the Feel Method. Analysis of the data, presented in Table 3, reveals the existence of two distinct soil textures: sandy clay loam, identified at the AR and EX sites, and clay loam, which is more widespread at the other sites.

Table 3: Physico-chemical properties of the soil samples studied

Site/land ownership	ZG	OD	LZ	OS	EH	LB	AR	EX
pH	7.8	7.6	7.4	7.7	7.7	7.6	7.7	8.1
SOM (%)	0.4	0.4	1.8	0.9	1	1.7	0.6	0.4
P2O5 (ppm)	20.3	37.4	100	78.6	36.3	72	24.1	23.3
K2O (ppm)	53	193.5	158.3	176	211	158.3	70.4	105.5
CaCO3T (%)	9.5	18.5	22	21	20	21	8	45
Salinity (g/kg)	0.1	0.2	0.7	1.1	0.1	0.9	0.3	1.2
Texture	Clay loam	Clay silt x	Clay silt x	Clay silt x	Clay loam	Clay loam	Sandy clay loam	Sandy clay loam

AR (Aarja); EX (Extension); EH (Elhammam); LB (Laabidate); LZ (Lamaiz); OD (Oudaghir); OS (Ouled slimane); ZG (Zenaga).

The soil properties measured, including both physical and chemical aspects, showed slight variations. The pH was slightly alkaline, varying between 7.4 and 8.1. The soil was predominantly limestone, with calcium carbonate ($CaCO_3$) levels ranging from 8% to 45%. In terms of organic matter, analyses revealed higher percentages at sites LZ and LB, with 1.8% and 1.7% respectively. On the other hand, the ZG, OD and EX sites recorded the lowest values, at 0.4% (see Table 3). Similarly, levels of phosphorus ($P\ O_{25}$) and potassium hydroxide (K2O) varied from site to site. The highest concentration of phosphorus was observed at site LZ, while site OD had the highest potassium value. In contrast, the highest Low levels of phosphorus and potassium hydroxide were noted at the ZG site. In terms of salinity, all samples showed moderate variations, ranging between 0.1 and 1.2 g/kg.

5.2 Number of MAC infectious propagules in soil

After one month of growing maize plants on serial dilutions, Table 4 shows the number of plants colonised by arbuscular mycorrhizal fungi (AMF) at different dilutions for the sites studied. It should be noted that all the soil samples from the different sites showed 100% mycorrhizal potential up to a dilution of 1/64. However, of all the concentrations tested, only the ZG and EX sites maintained this level of 100% mycorrhizal potential for all dilutions. At moderate to high dilutions (1/256 and 1/1024), MAC infection rates in soil samples from other sites, including OD, LZ, EH, LB and AR, ranged from 40% to 80%.

5.3 Root colonisation rate and spore density

The frequency and intensity of mycorrhization of Zea mays roots showed significant differences between sites after one month of cultivation ($p < 0.05$) (Figures 14 and 15). The ZG and EX sites stand out for their high frequency of mycorrhization, reaching 91% and 93%, respectively. In terms of colonisation intensity, the ZG and EX sites show a similar profile. On the other hand, the LZ site recorded the lowest value, with frequency and intensity percentages of no more than 39% and 9%, respectively. At the other sites, frequencies varied between 50 and 60%, and intensity levels between 10 and 20%, values that were not significantly different ($p > 0.05$).

Table 4: Number of maize plants showing traces of CMA at successive dilutions in soils sampled at the sites studied.

Repeat x5						
Dilution Site	1	1/4	1/16	1/64	1/256	1/1024
ZG	5	5	5	5	5	5
OD	5	5	5	5	3	3
LZ	5	5	5	5	3	2
OS	5	5	5	5	5	3
EH	5	5	5	5	4	3
LB	5	5	5	5	3	3
EX	5	5	5	5	5	5
AR	5	5	5	5	4	3

AR (Aarja); EX (Extension); EH (Elhammam); LB (Laabidate); LZ (Lamaiz); OD (Oudaghir); OS (Ouled slimane); ZG (Zenaga).

In terms of spore density, the soils at the ZG and EX sites stood out with a considerable number of mycorrhizal propagules, ranging from 25 to 28 per gram of soil, a quantity significantly different from that recorded at the other sites ($p < 0.05$) (Figure 16). In contrast, at site LZ, only 9 propagules per gram of soil were recorded. The other sites had a density of 11 to 13 propagules per gram of soil.

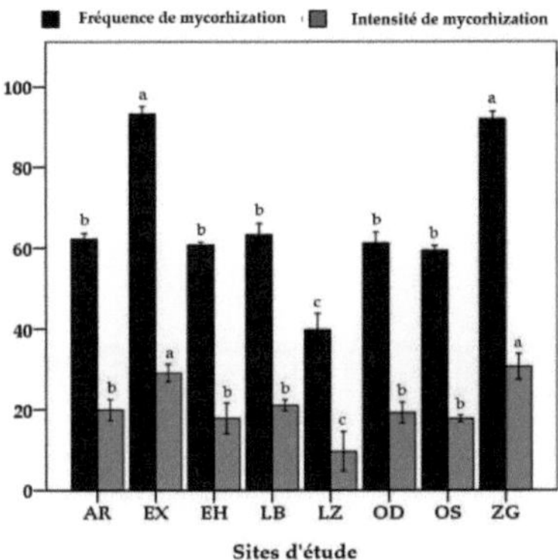

Figure 6: The frequency and intensity of mycorrhizal colonisation in Zea mays roots from eight study sites are presented. Sites include Aarja (AR); Extension (EX); Elhammam (EH); Laabidate (LB); Lamaiz (LZ); Oudaghir (OD); Ouled Slimane (OS); and Zenaga (ZG). The letters above the bars indicate statistically significant differences, determined by Tukey's test ($p < 0.05$).

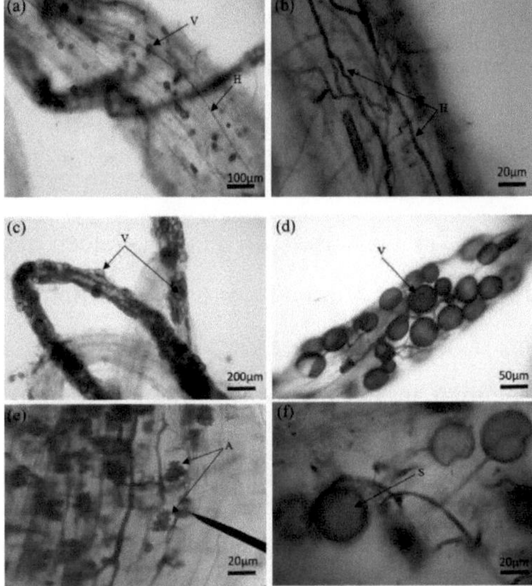

Figure 7: Structure of MAC colonisation in Zea mays trap plants. (a) colonisation by hyphae and vesicles; (b) colonisation by hyphae only; (c, d) colonisation by vesicles only; (e) colonisation by arbuscules; (f) intraradical spore.

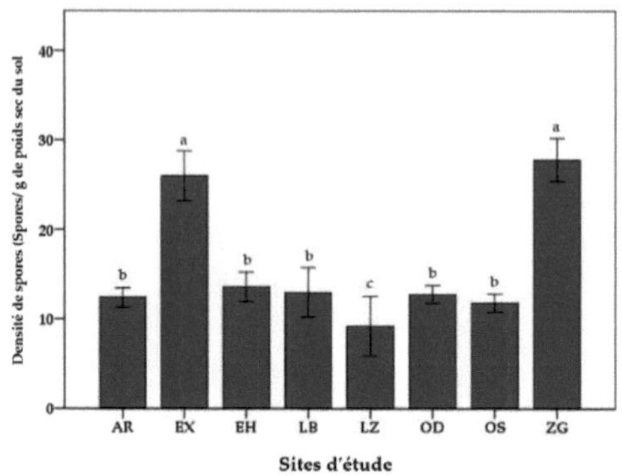

Figure 8: Spores density in rhizosphere soil samples from eight study sites: Aarja (AR), Extension (EX), Elhammam (EH), Laabidate (LB), Lamaiz (LZ), Oudaghir (OD), Ouled Slimane (OS), and Zenaga (ZG). The letters above the bars indicate significant differences according to the Tukey test ($p < 0.05$).

5.4 Correlations between CMA parameters and soil chemical characteristics

The Pearson coefficient (r) was used to quickly detect correlations between variables, with values ranging from -1 to 1. A coefficient of -1 indicates a negative correlation, 1 a positive correlation, and 0 the absence of linear correlation. The heat map in Figure 17 illustrates the correlation coefficients between the different MAC parameters in the rhizosphere of the date palm. These parameters include the frequency and intensity of mycorrhizal colonisation, spore density and soil chemical characteristics. Analyses revealed significant associations between these variables. In particular, mycorrhizal colonisation frequency was positively correlated with mycorrhization intensity and spore density, with correlation coefficients (r) of +0.931 and +0.923, respectively. Conversely, it was negatively correlated with soil phosphorus ($p = 0.005$), potassium ($p = 0.03$) and organic matter ($p = 0.04$) levels. Soil pH was positively associated with mycorrhizal frequency ($p = 0.02$) and spore density ($p = 0.007$), while spore density showed a negative correlation with phosphorus ($p= 0.006$) and potassium ($p = 0.007$), but positive with pH ($p = 0.006$). Finally, mycorrhization intensity was negatively correlated with phosphorus ($p = 0.007$) and potassium ($p = 0.008$) levels.

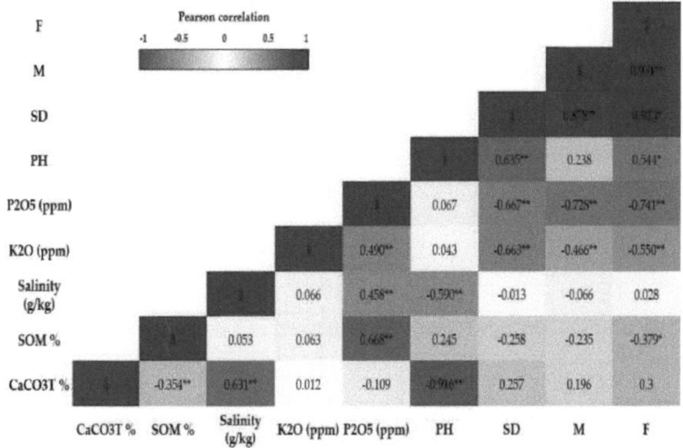

Figure 9: Heat map of the Pearson correlation coefficient between the chemical characteristics of the soil in the rhizosphere of Phoenix dactylifera L. and the parameters of the arbuscular mycorrhizal fungi (AMF). Mycorrhization frequency (F), mycorrhization intensity (M), spore density (SD). Correlation significance levels are indicated as follows: '*' for $p < 0.05$ and '**' for $p < 0.01$.

5.5 Morphological identification of CMAs

Soil samples collected from the rhizosphere of the Figuig oasis facilitated the isolation and identification of mycorrhizal spores. This analysis revealed the presence of 11 species in five genera: Rhizophagus, Funneliformis, Sclerocystis, Scutellospora and Acaulospora, detailed in Table 5 and Figure 18. Rhizophagus is the most dominant genus, with a prevalence of 70-90% and a high spore density, characterised by a multilayered wall fused with that of the underlying hyphae (see Figure 18d,e). The genera Scutellospora and Acaulospora make up between 10 and 20% of the samples. Acaulospora is characterised by spores that detach from a sporiferous saccule and subsequently become sessile, represented by a single species, Acaulospora sp., which is rarely observed (see Figure 18i). Finally, Funneliformis sp. and Rhizophagus sp. were found in similar proportions in all samples.

Table 5: CMA characteristics of surveyed soils

ID	Size	Colour	External wall	Inside wall	Suspension hyphae	Abondance	Shape	Internal structures	Number of walls	Spore alone or in a cluster
FIG 1	150 µm	Yellow-brown	Present	Present	Present	Mediocre	Globular	Droplets lipids	3	Visit cluster
FIG 2	120 µm	Brown	Present	Present	Present	Mediocre	Globular	Lipid droplets	2	Only
FIG 3	50 µm	Orange	Present	Present	Present	High	Spherical	Droplets lipids	2	Only
FIG 4	100 µm	Yellow-brown	Present	Absent	Present	Mediocre	Subglobular	Lipid droplets	2	In a cluster
FIG 5	300 µm	Orange yellow	Present	Present	Absent	High	Oval	Lipid droplets	3	In a cluster
FIG 6	100 µm	Yellow	Present	Present	Present	High	Subglobular	Lipid droplets	3	In a cluster
FIG 7	250 µm	Orange	Present	Absent	Absent	Rare	Globular	Lipid droplets	2	Only
FIG 8	130 µm	Red-brown	Present	Present	Present	Mediocre	Globular	Lipid droplets	3	Only
FIG 9	100 µm	Brown	Present	Present	Present	High	Oval	Lipid droplets	3	In a cluster
FIG 10	120 µm	Light yellow	Present	Present	Absent	Mediocre	Globular	Lipid droplets	2	Only
FIG 11	270 µm	Yellow	Present	Present	Present	Mediocre	Subglobular	Lipid droplets	2	In a cluster

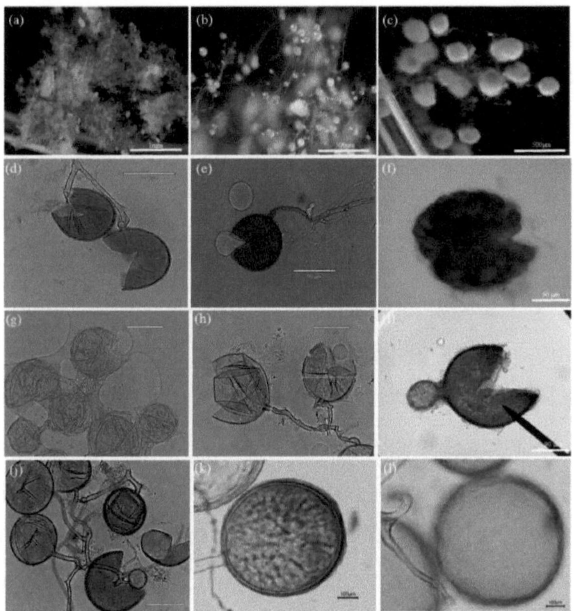

Figure 10: Numerous CMA spores were found in the rhizosphere of the Figuig oasis. Glomus sp. from pot cultures (a-c); Glomus sp. mounted in PVLG/Melzer (d,e); Sclerocystis sp. mounted in PVLG/Melzer (f); Glomus sp. mounted in PVLG (g,h); Acaulospora sp. mounted in PVLG (i); Glomus sp. mounted in PVLG (j,k); Scutellospora sp. mounted in PVLG (l).

5. Discussion

This study focused on assessing the mycorrhizal potential of the soil and identifying the AMCs in the rhizosphere of date palms in the Figuig oasis. It is essential to study mycorrhizal status in order to fully exploit the benefits of symbiosis between CMAs and plants. Faced with abiotic stress, plants deploy various strategies to increase their tolerance or to circumvent it. Previous studies [123,124] have revealed specific strategies that mitigate the negative impact of stress on plant growth. According to Oyediran et al. (2018) [125], plants in arid regions are able to produce large amounts of sugars and amino acids to withstand environmental stresses. In addition, the low phosphorus content of arid soils can favour symbiotic interactions between plants and fungi, thereby increasing the diversity of CMA spores in these ecosystems.Analysis of the correlation between the chemical parameters of the soil studied and the parameters of the mycorrhizal symbiosis (Figure 17) confirms this. Marschner and Cakmak (1986) [126] observed that the presence of certain chemicals in high concentrations tends to reduce the rate of mycorrhisation. Amijee et al (1989) [127] confirmed this trend for phosphorus, noting that root colonisation by mycorrhizae is highest in environments with low phosphorus concentrations and decreases with increasing concentration. Even very low phosphorus concentrations have been associated with a decrease in mycorrhization rates [124]. For example, at our study site LZ, where the phosphorus content is the highest (100 ppm), the frequency and intensity of mycorrhization are limited to around 39% and 9%, respectively. In contrast, at the ZG site with a phosphorus content of 20.3 ppm, the frequency and intensity of mycorrhization reached 91.99% and 30.56% respectively. Observations Similar results concerning reduced potassium and organic matter content were obtained at the ZG site. These results are in agreement with those reported by Oehl et al. (2010) [128]. Bhat et al. (2014) [129] also demonstrated that the availability of potassium and phosphorus in the soil and root colonisation by MACs may have a significant relationship. A significant positive correlation was observed between colonisation frequency and spore density in soil samples and the pH of our study sites, which ranged from 7.4 to 8.1. Toh et al. (2018) [130] reported that soil pH in the rhizosphere significantly affected spore number and root colonisation rate by MACs, a result in agreement with our findings. Several studies have shown a positive correlation between mycorrhizae and variations in soil pH [50,51]· Bainard et al. (2014) [131] found that some mycorrhizae species preferred acidic soils, while other studies indicate that variations in soil pH significantly influence the diversity of AMC populations [132]. However, Bainard et al. (2014) [131] also reported the lack of a significant correlation with soil pH, suggesting that variability in optimal pH values for different mycorrhizal species could explain this result. Melo et al. (2017) [133] highlighted this diversity, noting a negative correlation with pH for Acaulosporaceae and a positive correlation for indeterminate Glomoid. Other soil parameters, such as salinity and total limestone content, did not show a significant correlation, in line with the observations of Oehl et al (2010) [128]. Various genera of MAC spores were isolated and characterised from the soil samples, as shown in Table 5 and Figure 18. Characterisation focused on characteristics such as spore shape, spore colour, number of wall layers and other structures associated with MACs. Figure 18 shows that mycorrhizal fungi isolated from the soil belong to the genera Glomus sp., Acaulospora sp., Funneliformis sp., Rhizophagus sp., Sclerocystis sp. and Scutellospora sp., which are commonly found in arid and semi-arid habitats. Chebaane et al (2020) [134] reported the presence of several different isolates, including Funneliformis sp. and Rhizoglomus sp. in the rhizosphere of date palms in the Tunisian desert, a finding similar to that of Symanczik et al (2014) [135]. Glomus sp. Is the most abundant among the MAC species isolated from all samples, illustrating its significant prevalence in arid environments such as the Figuig oasis, the site of our study.

GENERAL CONCLUSION

This experiment highlighted the existence and diversity of arbuscular mycorrhizal fungi (AMF) within the rhizosphere of date palm soils in the Figuig oasis, underlining the importance of these results. Soil samples taken from the ZG and EX sites showed a higher percentage of isolated spores, indicating significant mycorrhizal activity in these specific areas.In addition, a negative correlation was observed between soil phosphorus content and the presence of these spores, confirming that mycorrhization is particularly effective in phosphorus-poor soils. In addition, a positive correlation was identified between soil pH and the presence of spores, suggesting that soil acidity or alkalinity has a favourable influence on MAC proliferation. These results confirm the agronomic interest of CMAs in soils deficient in essential nutrients, particularly phosphorus. The genus Glomus was the most abundant of the eight sites studied, highlighting its ability to adapt to the arid conditions of this region. In contrast, the genera Sclerocystis and Acaulospora were identified in very low proportions, suggesting a more restricted distribution or particular ecological requirements. In-depth study of the mycorrhizal potential of date palm rhizosphere soils, combined with the identification of MCAs, is of crucial importance for understanding and managing arid ecosystems such as the Figuig oasis. Through their symbiotic interaction with plant roots, these fungi significantly improve plant growth by optimising the uptake of essential nutrients, particularly phosphorus, which is often limited in the soils of this region. In this context, their role is fundamental to promote plant resilience in the face of environmental constraints such as water stress, salinity and soil poverty. The ability of CMAs to establish an effective symbiosis not only improves the nutritional status of palm trees, but also boosts their resistance to pathogens such as Fusarium oxysporum, and increases their tolerance to extreme climatic conditions. This mycorrhisation process thus contributes to the ecological stability of oases, where agriculture has to cope with limited water availability and often impoverished soils. The results of this study also highlight the importance of preserving and developing indigenous microbial biodiversity, such as CMAs, as these organisms are a valuable natural resource for the development of sustainable agricultural practices. By promoting crops adapted to local conditions, these fungi can reduce the use of chemical inputs (fertilisers and pesticides) and limit the ecological impact of farming activities. This approach is fully in line with the principles of agro-ecology and the sustainable management of natural resources, aimed at reconciling agricultural productivity with environmental protection. The knowledge gained from this research opens up new prospects for scientific exploration and agronomic innovation. It encourages further study of the interaction between CMAs and other plants in the region, leading to a better understanding of the dynamics of coevolution and the biological mechanisms behind this symbiosis. Future research could also explore the use of CMA as a bio-inoculant in reforestation or ecological restoration programmes, thereby helping to combat desertification and improve food security in arid areas. Ultimately, this study not only offers promising avenues for improving date palm cultivation, but also lays the foundations for the rational management of oasis ecosystems. Ultimately, a better understanding and targeted application of mycorrhizal symbioses could become essential levers for promoting innovative and resilient agricultural practices, while preserving the unique natural heritage of the Figuig oasis.

REFERENCES

1. Verner, D. Tunisia in a Changing Climate: Assessment and Actions for Increased Resilience and Development; World Bank Publications, 2013; ISBN 0-8213-9857-1.

2. Ghazouani, W.; Marlet, S.; Mekki, I.; Vidal, A. Farmers' Perceptions and Engineering Approach in the Modernization of a Community-managed Irrigation Scheme. A Case Study from an Oasis of the Nefzawa (South of Tunisia). Irrigation and Drainage **2009**, 58, S285-S296.

3. Hamza, H.; Jemni, M.; Benabderrahim, M.A.; Mrabet, A.; Touil, S.; Othmani, A.; Salah, M.B. Date Palm Status and Perspective in Tunisia. Date Palm Genetic Resources and Utilization: Volume 1: Africa and the Americas **2015**, 193-221.

4. Chao, C.T.; Krueger, R.R. The Date Palm (Phoenix Dactylifera L.): Overview of Biology, Uses, and Cultivation. horts **2007**, 42, 1077-1082, doi:10.21273/HORTSCI.42.5.1077.

5. Sharif, A.; Sanduk, M.; Taleb, H. The Date Palm and Its Role in Reducing Soil Salinity and Global Warming; 2010; pp. 59-64.

6. Rivera, D.; Obón, C.; Alcaraz, F.; Carreño, E.; Laguna, E.; Amorós, A.; Johnson, D.V.; Díaz, G.; Morte, A. Date Palm Status and Perspective in Spain. Date palm genetic resources and utilization: Volume 2: Asia and Europe **2015**, 489-526.

7. Habib, H.M.; Ibrahim, W.H. Nutritional Quality of 18 Date Fruit Varieties. International Journal of Food Sciences and Nutrition **2011**, 62, 544-551, doi:10.3109/09637486.2011.558073.

8. ANDZOA Synthèse Des Résultas de La Convention N° 03/CP/2012 Portant Sur l'étude Recensement, Caractérisation et Cartographie Des Palmeraies; 2019;

9. Killian, C.; Maire, R. Le Bayoud, Maladie Du Dattier; 1930;

10. Hakkou, A.; Bouakka, M. Oasis de Figuig: État Actuel de La Palmeraie et

Incidence de La Fusariose Vasculaire. Science et changements planétaires/Sécheresse **2004**, 15, 147-158.

11. Toutain, G.; Louvet, J. Lutte Contre Le Bayoud. IV. Orientations de La Lutte Au Maroc. Al-Awamia **1974**, 53, 114-162.

12. Louvet, J. and; Toutain, G. Recherches Sur Les Fusarioses. VII. Nouvelles Observations Sur La Fusariose Du Palmier Dattier et Précisions Concernant La Lutte; 1973.

13. Djerbi, M.; Aouad, L.; Filali, H.; Saaidi, M.; Chtioui, A.; Sedra, M.H.; Allaoui, M.; Hamdaoui, T.; Oubrich, M. Preliminary Results of Selection of High Quality Bayoud Resistant Clones among Natural Date Palm Population in Morocco; 1986; pp. 383- 399.

14. Pegna, F.G.; Sacchetti, P.; Canuti, V.; Trapani, S.; Bergesio, C.; Belcari, A.; Zanoni, B.; Meggiolaro, F. Radio Frequency Irradiation Treatment of Dates in a Single Layer to Control Carpophilus Hemipterus. Biosystems Engineering **2017**, 155, 1-11.

15. Abdelilah, M.; Ali, B. First Detection of Potosia Opaca Larva Attacks on Phoenix Dactylifera and Phoenix Canariensis in Morocco: Focus on Pests Control Strategies and Soil Quality of Prospected Palm Groves. Journal of Entomology and Zoology Studies **2017**, 5, 984-991.

16. Laouane, B.; Meddich, A.; Bechtaoui, N.; Oufdou, K.; Wahbi, S. Effects of Arbuscular Mycorrhizal Fungi and Rhizobia Symbiosis on the Tolerance of Medicago Sativa to Salt Stress. Gesunde Pflanzen **2019**, 71, 135-146.

17. Meddich, A.; Ait El Mokhtar, M.; Bourzik, W.; Mitsui, T.; Baslam, M.; Hafidi, M. Optimizing Growth and Tolerance of Date Palm (Phoenix Dactylifera L.) to Drought, Salinity, and Vascular Fusarium-Induced Wilt (Fusarium Oxysporum) by Application of Arbuscular Mycorrhizal Fungi (AMF). Root biology **2018**, 239-258.

18. Raklami, A.; Bechtaoui, N.; Tahiri, A.; Anli, M.; Meddich, A.; Oufdou, K. Use of Rhizobacteria and Mycorrhizae Consortium in the Open Field as a

Strategy for Improving Crop Nutrition, Productivity and Soil Fertility. Frontiers in Microbiology **2019**, 10, 1106.
19. Marschner, H.; Dell, B. Nutrient Uptake in Mycorrhizal Symbiosis. Plant and soil
1994, 159, 89-102.
20. Hashem, A.; Alqarawi, A.A.; Radhakrishnan, R.; Al-Arjani, A.-B.F.; Aldehaish, H.A.; Egamberdieva, D.; Abd_Allah, E.F. Arbuscular Mycorrhizal Fungi Regulate the Oxidative System, Hormones and Ionic Equilibrium to Trigger Salt Stress Tolerance in Cucumis Sativus L. Saudi journal of biological sciences **2018**, 25, 1102-1114.
21. Ait-El-Mokhtar, M.; Laouane, R.B.; Anli, M.; Boutasknit, A.; Wahbi, S.; Meddich, A. Use of Mycorrhizal Fungi in Improving Tolerance of the Date Palm (Phoenix Dactylifera L.) Seedlings to Salt Stress. Scientia Horticulturae **2019**, 253, 429-438.
22. Requena, N.; Perez-Solis, E.; Azcón-Aguilar, C.; Jeffries, P.; Barea, J.-M. Management of Indigenous Plant-Microbe Symbioses Aids Restoration of Desertified Ecosystems. Applied and environmental microbiology **2001**, 67, 495-498.
23. Huang, Y.-M.; Zou, Y.-N.; Wu, Q.-S. Alleviation of Drought Stress by Mycorrhizas Is Related to Increased Root H2O2 Efflux in Trifoliate Orange. Scientific reports **2017**, 7, 42335.
24. Estrada, B.; Aroca, R.; Azcón-Aguilar, C.; Barea, J.M.; Ruiz-Lozano, J.M. Importance of Native Arbuscular Mycorrhizal Inoculation in the Halophyte Asteriscus Maritimus for Successful Establishment and Growth under Saline Conditions. Plant and Soil **2013**, 370, 175-185.
25. Smith, S.E.; Read, D.J. Mycorrhizal Symbiosis; Academic press, 2010; ISBN 0-08- 055934-4.
26. Home | INVAM Available online: https://invam.ku.edu/ (accessed on 14 December 2023).

27. Błaszkowski, J. Glomeromycota; W. Szafer Institute of Botany, Polish Academy of Sciences, 2012; ISBN 83-89648-82-2.
28. Walker, C.; Cuenca, G.; Sánchez, F. Scutellospora Spinosissima Sp. Nov. a Newly Described Glomalean Fungus from Acidic, Low Nutrient Plant Communities in Venezuela. Annals of Botany **1998**, 82, 721-725.
29. Amf-Phylogeny.Com Available online: http://www.amf-phylogeny.com/ (accessed on 14 December 2023).
30. FIGUIG PROVINCIAL DEPARTMENT OF AGRICULTURE (DPA) (2009) Stratégies d'intervention de La DPA de Figuig; Ed. Ministère de l'Agriculture; p. 25 p;
31. Chafi, M. Problématique de l'eau Agricole Dans La Palmeraie de Figuig; 2007; pp. 5- 6.
32. TourerN, G. Le Palmier Dattier Culture et Production. Al awamia **1967**.
33. Sedra, M.H. Le Palm Dattier Base de La Mise En Valeur Des Oasis Au Maroc: Techniques Phoénicoles et Création d'oasis; INRA Editions, 2003; ISBN 9981-1994- 3-5.
34. El Hadrami, I.; El Bellaj, M.; El Idrissi, A.; J'Aiti, F.; El Jaafari, S.; Daayf, F. Biotechnologies Végétales et Amélioration Du Palm Dattier (Phoenix Dactylifera L.), Pivot de l'agriculture Oasienne Marocaine. Cahiers Agricultures **1998**, 7, 463-468.
35. Souna, F.; Himri, I.; Benabbas, R.; Fethi, F.; Chaib, C.; Bouakka, M.; Hakkou, A. Evaluation of Trichoderma Harzianum as a Biocontrol Agent against Vascular Fusariosis of Date Palm (Phoenix Dactylifera L.). Australian Journal of Basic and Applied Sciences **2012**, 6, 105-114.
36. Chakroune, K.; Bouakka, M.; Hakkou, A. Incidence de l'aération Sur Le Traitement Par Compostage Des Sous-Produits Du Palm Dattier Contaminés Par Fusarium Oxysporum f. Sp. Albedinis. Canadian journal of microbiology **2005**, 51, 69-77.
37. Toutain, G. Observations on the Spread of an Active Focus of Bayoud

Disease in a Plantation of Regularly Spaced Date Palms. Awamia **1970**, 155-160.

38. Sedra, M.H. Screening of a Collection of Date Palm Genotypes for Resistance to Bayoud Caused by Fusarium Oxysporum f. Sp. Albedinis. Al Awamia **1995**, 90, 9-18.
39. Saaidi, M. Comportement Au Champ de 32 Cultivars de Palm Dattier Vis-à-Vis Du Bayoud: 25 Années d'observations. Agronomie **1992**, 12, 359-370.
40. Louvet, J.; Bulit, J.; Toutain, G.; Rieuf, P. LE BAYOUD, FUSARIOSE VASCULAIRE DU PALMIER DATTIER SYMPTOMS ET NATURE DT] THE DISEASE MEANS OF CONTROL: Ë. Al Awamia **1970**, 35, 161-162.
41. Jilali, A. Impact of Climate Change on the Figuig Aquifer Using a Numerical Model: Oasis of Eastern Morocco. Journal of Biology and Earth Sciences **2014**, 4, E16-E24.
42. Schilling, J.; Freier, K.P.; Hertig, E.; Scheffran, J. Climate Change, Vulnerability and Adaptation in North Africa with Focus on Morocco. Agriculture, Ecosystems & Environment **2012**, 156, 12-26.
43. Szabolcs, I. Review of Research on Salt-Affected Soils. Soil Science **1981**, 131, 63.
44. Hayward, H.E. Plant Growth under Saline Conditions; Unesco, 1952;
45. Dubost, D.; Moguedet, G. La Révolution Hydraulique Dans Les Oasis Impose Une Nouvelle Gestion de l'eau Dans Les Zones Urbaines. Méditerranée **2002**, 99, 15-20.
46. Dubost, D. Pratique de l'irrigation Au Sahara. CIHEAM/IAM **1994**.
47. Ayers, R.S.; Westcot, D.W. Water Quality for Agriculture; Food and Agriculture Organization of the United Nations Rome, 1985; Vol. 29; ISBN 92-5-102263-1.
48. Daoud, Y.; Halitim, A. Irrigation et Salinisation Au Sahara Algérien. Science et changements planétaires/Sécheresse **1994**, 5, 151-160.
49. Les Sols Des Zones Oasiennes En Proie à Une Baisse de Fertilité - Médias24

Available online: https://medias24.com/2023/05/31/les-sols-des-zones-oasiennes-en-proie-a-une- baisse-de-fertilite/ (accessed on 21 December 2023).
50. La Dégradation Des Sols En France et Dans Le Monde, Une Catastrophe Écologique Ignorée | Planet-Vie Available online: https://planet-vie.ens.fr/thematiques/ecologie/gestion-de-l-environnement-pollution/la-degradation- des-sols-en-france-et-dans (accessed on 21 December 2023).
51. GLO French_Ch9.Pdf.
52. CNRS/Sagascience - Agricultural Management Methods and Influences on Soil Biodiversity Available online: https://www.cnrs.fr/cw/dossiers/dosbiodiv/index.php?pid=decouv_chapC_p5_d1&zoo m_id=zoom_d1_2 (accessed on 22 December 2023).
53. Fiche-Actu-Oasis-2021.Pdf.
54. GLO French_Ch9.Pdf.
55. Claire Horner-Devine, M.; Leibold, M.A.; Smith, V.H.; Bohannan, B.J. Bacterial Diversity Patterns along a Gradient of Primary Productivity. Ecology letters **2003**, 6, 613-622.
56. Curtis, T.P.; Sloan, W.T.; Scannell, J.W. Estimating Prokaryotic Diversity and Its Limits. Proceedings of the National Academy of Sciences **2002**, 99, 10494-10499.
57. Leake, J.; Johnson, D.; Donnelly, D.; Muckle, G.; Boddy, L.; Read, D. Networks of Power and Influence: The Role of Mycorrhizal Mycelium in Controlling Plant Growth. Communities and Agroecosystem Functioning. Canadian Journal of Botany **2004**, 82, 1016-1045.
58. Kowalchuk, G.A.; Stephen, J.R. Ammonia-Oxidizing Bacteria: A Model for Molecular Microbial Ecology. Annual Reviews in Microbiology **2001**, 55, 485-529.
59. N Högberg, M.; Högberg, P. Extramatrical Ectomycorrhizal Mycelium Contributes One-Third of Microbial Biomass and Produces, Together with Associated Roots, Half the Dissolved Organic Carbon in a Forest Soil. New

Phytologist **2002**, 154.
60. Whittaker, R.H. New Concepts of Kingdoms of Organisms: Evolutionary Relations Are Better Represented by New Classifications than by the Traditional Two Kingdoms. Science **1969**, 163, 150-160.
61. Peyret-Guzzon, M.P. Etudes Moléculaires de La Diversité Des Communautés et Populations de Champignons Mycorhiziens à Arbuscules (Glomeromycota). **2014**.
62. Wainright, P.O.; Hinkle, G.; Sogin, M.L.; Stickel, S.K. Monophyletic Origins of the Metazoa: An Evolutionary Link with Fungi. Science **1993**, 260, 340-342.
63. Baldauf, S.L.; Palmer, J.D. Animals and Fungi Are Each Other's Closest Relatives: Congruent Evidence from Multiple Proteins. Proceedings of the National Academy of Sciences **1993**, 90, 11558-11562.
64. Baldauf, S.L. A Search for the Origins of Animals and Fungi: Comparing and Combining Molecular Data. the american naturalist **1999**, 154, 178-188.
65. Latgé, J. Tasting the Fungal Cell Wall. Cellular microbiology **2010**, 12, 863-872.
66. Blackwell, M. The Fungi: 1, 2, 3... 5.1 Million Species? American journal of botany **2011**, 98, 426-438.
67. Hibbett, D.S.; Binder, M.; Bischoff, J.F.; Blackwell, M.; Cannon, P.F.; Eriksson, O.E.; Huhndorf, S.; James, T.; Kirk, P.M.; Lücking, R. A Higher-Level Phylogenetic Classification of the Fungi. Mycological research **2007**, 111, 509-547.
68. Ahmadjian, V. The Lichen Photobiont: What Can It Tell Us about Lichen Systematics? Bryologist **1993**, 310-313.
69. Van Der Heijden, M.G.; Streitwolf-Engel, R.; Riedl, R.; Siegrist, S.; Neudecker, A.; Ineichen, K.; Boller, T.; Wiemken, A.; Sanders, I.R. The Mycorrhizal Contribution to Plant Productivity, Plant Nutrition and Soil Structure in Experimental Grassland. New phytologist **2006**, 172, 739-752.

70. Brundrett, M.C.; Tedersoo, L. Evolutionary History of Mycorrhizal Symbioses and Global Host Plant Diversity. New Phytologist **2018**, 220, 1108-1115.
71. Redecker, D. Specific PCR Primers to Identify Arbuscular Mycorrhizal Fungi within Colonized Roots. Mycorrhiza **2000**, 10, 73-80.
72. Schüssler, A.; Walker, C. The Glomeromycota: A Species List With New Families and New Gener. The Glomeromycota: A Species List With New Families and New Gener **2010**.
73. Redecker, D.; Schüßler, A.; Stockinger, H.; Stürmer, S.L.; Morton, J.B.; Walker, C. An Evidence-Based Consensus for the Classification of Arbuscular Mycorrhizal Fungi (Glomeromycota). Mycorrhiza **2013**, 23, 515-531.
74. Parniske, M. Arbuscular Mycorrhiza: The Mother of Plant Root Endosymbioses.
Nature Reviews Microbiology **2008**, 6, 763-775.
75. Keymer, A.; Pimprikar, P.; Wewer, V.; Huber, C.; Brands, M.; Bucerius, S.L.; Delaux, P.-M.; Klingl, V.; Röpenack-Lahaye, E. von; Wang, T.L. Lipid Transfer from Plants to Arbuscular Mycorrhiza Fungi. elife **2017**, 6, e29107.
76. SCHÜßLER, A.; Schwarzott, D.; Walker, C. A New Fungal Phylum, the Glomeromycota: Phylogeny and Evolution. Mycological research **2001**, 105, 1413-1421.
77. Oehl, F.; Sieverding, E.; Palenzuela, J.; Ineichen, K.; Da Silva, G.A. Advances in Glomeromycota Taxonomy and Classification. IMA Fungus **2011**, 2, 191-199, doi:10.5598/imafungus.2011.02.02.10.
78. Öpik, M.; Zobel, M.; Cantero, J.J.; Davison, J.; Facelli, J.M.; Hiiesalu, I.; Jairus, T.; Kalwij, J.M.; Koorem, K.; Leal, M.E.; et al. Global Sampling of Plant Roots Expands the Described Molecular Diversity of Arbuscular Mycorrhizal Fungi. Mycorrhiza **2013**, 23, 411-430, doi:10.1007/s00572-013-0482-2.
79. Besserer, A.; Puech-Pagès, V.; Kiefer, P.; Gomez-Roldan, V.; Jauneau, A.; Roy, S.; Portais, J.-C.; Roux, C.; Bécard, G.; Séjalon-Delmas, N. Strigolactones

Stimulate Arbuscular Mycorrhizal Fungi by Activating Mitochondria. PLoS biology **2006**, 4, e226.

80. Gianinazzi-Pearson, V.; Branzanti, B.; Gianinazzi, S. In Vitro Enhancement of Spore Germination and Early Hyphal Growth of a Vesicular-Arbuscular Mycorrhizal Fungus by Host Root Exudates and Plant Flavonoids. Symbiosis **1989**.

81. Buee, M.; Rossignol, M.; Jauneau, A.; Ranjeva, R.; Bécard, G. The Pre-Symbiotic Growth of Arbuscular Mycorrhizal Fungi Is Induced by a Branching Factor Partially Purified from Plant Root Exudates. Molecular Plant-Microbe Interactions **2000**, 13, 693-698.

82. Delaux, P.; Bécard, G.; Combier, J. NSP 1 Is a Component of the Myc Signaling Pathway. New Phytologist **2013**, 199, 59-65.

83. Genre, A.; Chabaud, M.; Balzergue, C.; Puech-Pagès, V.; Novero, M.; Rey, T.; Fournier, J.; Rochange, S.; Bécard, G.; Bonfante, P. Short-chain Chitin Oligomers from Arbuscular Mycorrhizal Fungi Trigger Nuclear C A2+ Spiking in M Edicago Truncatula Roots and Their Production Is Enhanced by Strigolactone. New Phytologist **2013**, 198, 190-202.

84. Oláh, B.; Brière, C.; Bécard, G.; Dénarié, J.; Gough, C. Nod Factors and a Diffusible Factor from Arbuscular Mycorrhizal Fungi Stimulate Lateral Root Formation in Medicago Truncatula via the DMI1/DMI2 Signalling Pathway. The Plant Journal **2005**, 44, 195-207.

85. Sharif, A.O.; Sanduk, M.; Taleb, H.M. THE DATE PALM AND ITS ROLE IN REDUCING SOIL SALINITY AND GLOBAL WARMING. Acta Hortic. **2010**, 59-64, doi:10.17660/ActaHortic.2010.882.5.

86. Rivera, D.; Obón, C.; Alcaraz, F.; Carreño, E.; Laguna, E.; Amorós, A.; Johnson, D.V.; Díaz, G.; Morte, A. Date Palm Status and Perspective in Spain. In Date Palm Genetic Resources and Utilization; Al-Khayri, J.M., Jain, S.M., Johnson, D.V., Eds; Springer Netherlands: Dordrecht, 2015; pp. 489-526 ISBN 978-94-017-9706-1.

87. Ghazouani, W.; Marlet, S.; Mekki, I.; Vidal, A. Farmers' Perceptions and Engineering Approach in the Modernization of a Community-Managed Irrigation Scheme. A Case Study from an Oasis of the Nefzawa (South of Tunisia). Irrig. and Drain. **2009**, 58, S285-S296, doi:10.1002/ird.528.
88. Hamed, Y.; Hadji, R.; Redhaounia, B.; Zighmi, K.; Bâali, F.; El Gayar, A. Climate Impact on Surface and Groundwater in North Africa: A Global Synthesis of Findings and Recommendations. Euro-Mediterr J Environ Integr **2018**, 3, 25, doi:10.1007/s41207-018-0067-8.
89. Haj-Amor, Z.; Tóth, T.; Ibrahimi, M.-K.; Bouri, S. Effects of Excessive Irrigation of Date Palm on Soil Salinization, Shallow Groundwater Properties, and Water Use in a Saharan Oasis. Environ Earth Sci **2017**, 76, 590, doi:10.1007/s12665-017-6935-8.
90. Hachicha, M.; Ben Aissa, I. Managing Salinity in Tunisian Oases. Journal of Life Sciences **2014**, 8, 775-782.
91. Zeddouk, M. La Problématique Du Développement Agricole Dans Le Milieu Oasien: Cas Du Tafilalet. In Proceedings of the Actes du Symposium International sur le Développement Durable des Systèmes Oasiens du; 2005; Vol. 8, pp. 635-645.
92. BOUAMMAR, B. Le Développement Agricole Dans Les Régions Sahariennes Etude de Cas de La Région de Ouargla et de La Région de Biskra (2006-2008), 2010.
93. Abohatem, M.; Zouine, J.; El Hadrami, I. Low Concentrations of BAP and High Rate of Subcultures Improve the Establishment and Multiplication of Somatic Embryos in Date Palm Suspension Cultures by Limiting Oxidative Browning Associated with High Levels of Total Phenols and Peroxidase Activities. Scientia Horticulturae **2011**, 130, 344-348, doi:10.1016/j.scienta.2011.06.045.
94. Newsham, K.K.; Fitter, A.H.; Watkinson, A.R. Arbuscular Mycorrhiza Protect an Annual Grass from Root Pathogenic Fungi in the Field. Journal of

ecology **1995**, 991- 1000.

95. Smith, S.E.; Jakobsen, I.; Grønlund, M.; Smith, F.A. Roles of Arbuscular Mycorrhizas in Plant Phosphorus Nutrition: Interactions between Pathways of Phosphorus Uptake in Arbuscular Mycorrhizal Roots Have Important Implications for Understanding and Manipulating Plant Phosphorus Acquisition. Plant physiology **2011**, 156, 1050-1057.

96. Augé, R.M. Water Relations, Drought and Vesicular-Arbuscular Mycorrhizal Symbiosis. Mycorrhiza **2001**, 11, 3-42.

97. Neumann, E.; George, E. Colonisation with the Arbuscular Mycorrhizal Fungus Glomus Mosseae (Nicol. & Gerd.) Enhanced Phosphorus Uptake from Dry Soil in Sorghum Bicolor (L.). Plant and Soil **2004**, 261, 245-255.

98. Rillig, M.C.; Mummey, D.L. Mycorrhizas and Soil Structure. New phytologist **2006**,171, 41-53.

99. Cui, M.; Nobel, P.S. Nutrient Status, Water Uptake and Gas Exchange for Three Desert Succulents Infected with Mycorrhizal Fungi. New Phytologist **1992**, 122, 643-649.

100. Meddich, A.; Jaiti, F.; Bourzik, W.; El Asli, A.; Hafidi, M. Use of Mycorrhizal Fungi as a Strategy for Improving the Drought Tolerance in Date Palm (Phoenix Dactylifera). Scientia Horticulturae **2015**, 192, 468-474.

101. Duponnois, R.; Ramanankierana, H.; Hafidi, M.; Baohanta, R.; Baudoin, E.; Thioulouse, J.; Sanguin, H.; Ba, A.; Galiana, A.; Bally, R. Native Plant Resources to Optimize the Performances of Forest Rehabilitation in Mediterranean and Tropical Environment: Some Examples of Nursing Plant Species That Improve the Soil Mycorrhizal Potential. Comptes Rendus Biologies **2013**, 336, 265-272.

102. Marulanda, A.; Porcel, R.; Barea, J.M.; Azcón, R. Drought Tolerance and Antioxidant Activities in Lavender Plants Colonized by Native Drought-Tolerant or Drought- Sensitive Glomus Species. Microbial ecology **2007**, 54, 543-552.

103. Lekberg, Y.; Koide, R.T.; Rohr, J.R.; ALDRICH-WOLFE, L.; Morton, J.B. Role of Niche Restrictions and Dispersal in the Composition of Arbuscular Mycorrhizal Fungal Communities. Journal of Ecology **2007**, 95, 95-105.
104. Antunes, P.M.; Koch, A.M.; Morton, J.B.; Rillig, M.C.; Klironomos, J.N. Evidence for Functional Divergence in Arbuscular Mycorrhizal Fungi from Contrasting Climatic Origins. New Phytologist **2011**, 189, 507-514, doi:10.1111/j.1469-8137.2010.03480.x.
105. Eaton, A.D.; Franson, M.A.H.; Clesceri, L.S.; Rice, E.W.; Greenberg, A.E. Standard Methods for the Examination of Water & Wastewater. In Standard methods for the examination of water & wastewater; 2005; p. 1. v-1. v.
106. Olsen, S.R. Estimation of Available Phosphorus in Soils by Extraction with Sodium Bicarbonate; US Department of Agriculture, 1954;
107. Mathieu, C.; Pieltain, F.; Jeanroy, E. Analyse Chimique Des Sols: Méthodes Choisies; Tec & doc, 2003; ISBN 2-7430-0620-X.
108. Mathieu, C.; Pieltain, F. Chemical Soil Analysis: Chosen Methods. Chemical soil analysis: chosen methods. **2003**.
109. He, Y.; DeSutter, T.; Prunty, L.; Hopkins, D.; Jia, X.; Wysocki, D.A. Evaluation of 1:5 Soil to Water Extract Electrical Conductivity Methods. Geoderma **2012**, 185-186, 12- 17, doi:10.1016/j.geoderma.2012.03.022.
110. NF P94-048 Soil: Investigation and Testing - Determination of the Carbonate Content - Calcimeter Method 2002.
111. Ritchey, E.L.; McGrath, J.M.; Gehring, D. Determining Soil Texture by Feel. **2015**.
112. Plenchette, C.; Perrin, R.; Duvert, P. The Concept of Soil Infectivity and a Method for Its Determination as Applied to Endomycorrhizas. Canadian Journal of Botany **1989**, 67, 112-115.
113. Liu, R.; Wang, F. Selection of Appropriate Host Plants Used in Trap Culture of Arbuscular Mycorrhizal Fungi. Mycorrhiza **2003**, 13, 123-127.
114. Phillips, J.M.; Hayman, D.S. Improved Procedures for Clearing Roots and

Staining Parasitic and Vesicular-Arbuscular Mycorrhizal Fungi for Rapid Assessment of Infection. Transactions of the British Mycological Society **1970**, 55, 158-IN18, doi:10.1016/S0007-1536(70)80110-3.

115. Fisher, R.A.; Yates, F. Statistical Tables for Research Workers. Oliver and Boyd: London **1949**.

116. Gerdemann, J.W.; Nicolson, T.H. Spores of Mycorrhizal Endogone Species Extracted from Soil by Wet Sieving and Decanting. Transactions of the British Mycological Society **1963**, 46, 235-244, doi:10.1016/S0007-1536(63)80079-0.

117. Koskey, R.E. A Convenient, Permanent Slide Mounting Medium. Newsletter Mycol. Soc. Amer. **1983**, 34, 59.

118. Brundrett, M. Practical Methods in Mycorrhiza Research: Based on a Workshop Organized in Conjunction with the Ninth North American Conference on Mycorrhizae, University of Guelph, Guelph, Ontario, Canada (No Title) **1994**.

119. Http://Www.Amf-Phylogeny.Com/.

120. Koske, R.E.; Gemma, J.N. A Modified Procedure for Staining Roots to Detect VA Mycorrhizas. Mycological Research **1989**, 92, 486-488, doi:10.1016/S0953-
7562(89)80195-9.

121. Trouvelot, A. Measure Du Taux de Mycorrhization d'un Systeme Radiculaire. Recherche de Methods d'estimation Ayant Une Signification Fonctionnelle. Physiological and genetical aspects of mycorrhizae **1986**, 217-221.Introduction Available online: https://www2.dijon.inrae.fr/mychintec/Mycocalc- prg/download.html (accessed on 26 December 2023).

122. Evelin, H.; Kapoor, R.; Giri, B. Arbuscular Mycorrhizal Fungi in Alleviation of Salt Stress: A Review. Annals of botany **2009**, 104, 1263-1280.

123. Saharan,B.S.; Nehra, V. Plant Growth Promoting Rhizobacteria: A Critical Review.Life Sci Med Res **2011**, 21, 30.

124. Oyediran, O.K.; Kumar, A.G.; Neelam, J. Arbuscular Mycorrhizal Fungi Associated with Rhizosphere of Tomato Grown in Arid and Semi-Arid Regions of Indian Desert. Asian Journal of Agricultural Research **2018**, 12, 10-18.

125. Marschner, H.; Cakmak, I. Mechanism of Phosphorus-induced Zinc Deficiency in Cotton. II. Evidence for Impaired Shoot Control of Phosphorus Uptake and Translocation under Zinc Deficiency. Physiologia plantarum **1986**, 68, 491-496.

126. KAPOOR, R.; GIRI, B.; MUKERJI, K. Occurrence of Vesicular Arbuscular Mycorrhiza. Techniques in Mycorrhizal Studies **2013**, 51.

127. Oehl, F.; Laczko, E.; Bogenrieder, A.; Stahr, K.; Bösch, R.; van der Heijden, M.; Sieverding, E. Soil Type and Land Use Intensity Determine the Composition of Arbuscular Mycorrhizal Fungal Communities. Soil Biology and Biochemistry **2010**, 42, 724-738.

128. BHAT, B.A.; SHEIKH, M.A.; TIWARI, A. J Presearch ARTICLE. International Journal of Plant Sciences **2014**, 9, 1-6.

129. Toh, S.; Lihan, S.; Yong, C.; Tiang, B.; Rakiya, A.; Edward, R. Isolation and Characterisation of Arbuscular Mycorrhizal (AM) Fungi Spores from Selected Plant Roots and Their Rhizosphere Soil Environment. Malaysian Journal of Microbiology **2018**, 14, 335-343.

130. Bainard, L.D.; Bainard, J.D.; Hamel, C.; Gan, Y. Spatial and Temporal Structuring of Arbuscular Mycorrhizal Communities Is Differentially Influenced by Abiotic Factors and Host Crop in a Semi-Arid Prairie Agroecosystem. FEMS Microbiology Ecology **2014**, 88, 333-344.

131. Mosbah, M.; Philippe, D.L.; Mohamed, M. Molecular Identification of Arbuscular Mycorrhizal Fungal Spores Associated to the Rhizosphere of Retama Raetam in Tunisia. Soil Science and Plant Nutrition **2018**, 64, 335-341.

132. Melo, C.D.; Luna, S.; Krüger, C.; Walker, C.; Mendonça, D.; Fonseca, H.M.; Jaizme- Vega, M.; da Câmara Machado, A. Arbuscular Mycorrhizal Fungal Community Composition Associated with Juniperus Brevifolia in Native

Azorean Forest. Acta Oecologica **2017**, 79, 48-61.
133. Chebaane, A.; Symanczik, S.; Oehl, F.; Azri, R.; Gargouri, M.; Mäder, P.; Mliki, A.; Fki, L. Arbuscular Mycorrhizal Fungi Associated with Phoenix Dactylifera L. Grown in Tunisian Sahara Oases of Different Salinity Levels. Symbiosis **2020**, 81, 173-186.
134. Symanczik, S.; Błaszkowski, J.; Koegel, S.; Boller, T.; Wiemken, A.; Al-Yahya'Ei, M.N. Isolation and Identification of Desert Habituated Arbuscular Mycorrhizal Fungi Newly Reported from the Arabian Peninsula. Journal of Arid Land **2014**, 6, 488-497.

Printed by Books on Demand GmbH, Norderstedt / Germany